Matheus Carvalho
Pedro Lucas
Leonardo Monteiro

Practical Studies - Applied Mechanical Engineering

Matheus Carvalho
Pedro Lucas
Leonardo Monteiro

Practical Studies - Applied Mechanical Engineering

Mechanical Engineering Applications

ScienciaScripts

Imprint

Cover image: www.ingimage.com

This book is a translation from the original published under ISBN 978-620-6-75774-0.

Publisher:
Sciencia Scripts
is a trademark of
Dodo Books Indian Ocean Ltd. and OmniScriptum S.R.L publishing group

120 High Road, East Finchley, London, N2 9ED, United Kingdom
Str. Armeneasca 28/1, office 1, Chisinau MD-2012, Republic of Moldova, Europe
Printed at: see last page
ISBN: 978-620-3-63329-0

Contents

CHAPTER 1

SUMMARY

With the universal use of air conditioning indoors, over the years it has been realised how much it facilitates personal relationships and, consequently, business relationships such as sales, for example. The use of air conditioning systems aims to provide thermal comfort and improve indoor air quality. In commercial environments, the presence of an air-conditioning system brings greater quality to the employees and customers who frequent them by controlling the indoor temperature. As a preliminary step to sizing the air conditioning system, it is necessary to study the thermal load of the shop under study. Therefore, this work presents a study of the thermal load of the environment in question, with the aim of helping with a subsequent air conditioning project. The methodology used in this work will be based on the regulatory standard ABNT NBR 16401, the ASHRAE methodology and the use of software in a BIM environment, which promotes greater assertiveness in the data obtained, as well as facilitating the analysis of the room's thermal load.

Keywords: BIM environment; Air conditioning; Thermal load; Thermal comfort.

1 INTRODUCTION

The knowledge and resources of refrigeration have existed in society for hundreds of years. The search for physical well-being dates back to the earliest times, when people used ice to preserve food for later consumption. However, little by little the initial perspectives on these resources were transformed into another possibility: promoting thermal comfort in environments with adverse climates. With this advent, in recent years scholars and specialists have sought to refine studies on the effects of thermal comfort in enclosed spaces, because according to Silva (2005) the implementation of air conditioning has indirectly led to greater employee productivity due to the increase in the body's fungi when subjected to thermally comfortable environments.

Various elements influence the conditions of an internal environment, including: the characteristics of the external environment, the construction details of the building, the routine of its inhabitants and the activities they carry out in the space. Each building has specific variables that require monitoring and behaves differently when compared to other buildings. Knowledge of the particularities of the premises, especially those relating to air quality, is extremely important. This is because air plays a crucial role in the interaction between the climate, the building and its occupants. In addition, according to Karashima (2006), accounting for all the factors mentioned above makes the thermal load calculation more detailed. Furthermore, guaranteeing the quality of indoor air in the building is of paramount importance so that occupants are protected and safe from the harmful effects of certain contaminants (whether of physical, chemical or biological origin) (MORENO, 2017).

1.1 BACKGROUND

Representing 11.04% of the Gross Domestic Product (GDP) and generating more than 1.2 million formal jobs in December 2020 (SBVC, 2021), the retail sector is one of the most important for the Brazilian economy. To contribute to this growth, it is necessary for shops to be constantly adapting to the market, offering benefits to employees and customers, such as acoustic and thermal comfort.

That said, it is essential to take into account geographical and meteorological factors, given that Brazil's climate changes at specific times throughout the year, causing problems in certain regions due to thermal sensations, especially in the north-east of the country.

Located on the northeastern coast of Brazil, Recife registers high temperatures throughout the winter period, especially in June. According to the National Institute of Meteorology (INMET, 2021), Recife recorded the highest heat value of the period in June 2021, reaching a maximum temperature of 32°C, making it a high temperature when analysing the season.

Given the problem of thermal sensations in the city, it is necessary for commercial environments with the presence of individuals to be equipped with an air conditioning system. Therefore, it is essential to design an air conditioning system that calculates and estimates in its planning the various factors that may have an impact on the heat generated in the room. Therefore, prior to sizing, it is essential to obtain the thermal load, using software in the BIM *(Building Intelligence Modelling)* environment, so that an efficient engineering project can be carried out, given that these programmes aim to focus on the project's life cycle (MARCOS, 2009).

1.2 PROBLEMATICS

The company in the following study is seeking to improve its standard of customer service, given that it is planning to set up in all the states of north-eastern Brazil. Also in this context of expansion and market recognition, the company is looking to reformulate its layout, altering the structural plant and installing an air conditioning system, because in the context of retail modernisation, the physical environment is now recognised as a fundamental variable for the image of the business (VAROTTO, 2018).

In this way, the BIM flow presents the most viable solution for subsequent implementation, ensuring a higher quality of information for the sizing of the building's air conditioning system.

1.3 OBJECTIVES

The objectives of this study were segmented into general and specific in order to generate a better understanding of the methodology.

1.3.1 General Objective

The general objective is to obtain the shop's thermal load in a BIM environment, covering all the factors and situations to be analysed.

1.3.2 Specific Objectives

- Simplification and visual analysis of the shop structure together with the floor plan in CAD *(ComputerAided Design);*

Creation of an IFC model using the shop floor plan in CAD;

Designing the problem and defining the control conditions;

Research into room heating hypotheses.

2 THEORETICAL BASIS

2.1 Fundamental Concepts

In order to study thermal load in detail, it is essential to understand some fundamental concepts, which dictate the fundamentals of the 1st law of thermodynamics and the fundamentals of heat transfer (conduction, convection and radiation).

These definitions, according to Marques (2006), have the following concepts:

- Thermodynamic properties: these are all the macroscopic characteristics of a given system: pressure, temperature, volume, etc;
- Thermodynamic state: understood as the condition in which a substance finds itself, characterised by its properties;
- Thermodynamic process: change in the state of a system. A process is any change in the properties of a given system;
- Cycle: defined as a series of processes in which the initial and final states of the system coincide;
- Pure substance: any substance whose chemical composition is homogeneous and invariable;
- Saturation temperature: defined as the temperature at which a pure substance vaporises at a given pressure;
- Saturated liquid: when a given substance becomes saturated.

finds itself as a liquid at saturation temperature and pressure;

- Undercooled liquid: a liquid is considered to be undercooled when its temperature is lower than its saturation temperature for the existing pressure;
- Title (X): When a substance is part liquid and

part vapour, at saturation temperature, the ratio between the mass of vapour and the total mass, i.e. the mass of liquid plus the mass of vapour, is called the titre (x);

$$X = \frac{Mv}{M1+Mv} = \frac{Mv}{Mt} \qquad (2.1)$$

- Saturated vapour: this is given when the substance is completely vapour at the saturation temperature, and its titer is given as 1 or 100%;
- Control volume: The size and shape of the volume can be defined in such a way as to simplify the analysis to be carried out. The surface of the control volume can be fixed or movable, but its movement must be referenced to some coordinate system (WYLEN, 2003).

2.2 PRINCIPLES OF HEAT TRANSFER

Before starting to calculate the thermal load of a room, it is essential to know the principles governing heat transfer.

Heat is thermal energy in transit due to a temperature difference in space (INCROPERA, 2008), which occurs in three ways: conduction, convection and radiation. The following figure shows the three types of heat transfer.

Figure 1 - Heat transfer mechanisms.

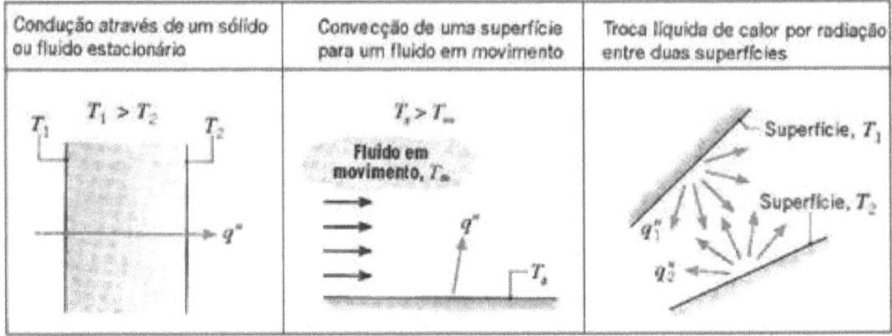

Source: INCROPERA (2008).

2.2.1 Heat Transfer by Conduction

When there is a temperature gradient in a stationary medium, which can be a solid or fluid, heat transfer occurs by conduction. Conduction can be seen as the transfer of energy from the more energetic to the less energetic particles of a substance due to interactions between particles (INCROPERA, 2014). Figure 3 shows two types of surfaces transferring heat by conduction.

Figure 2 - Heat transfer by conduction on two types of surface.

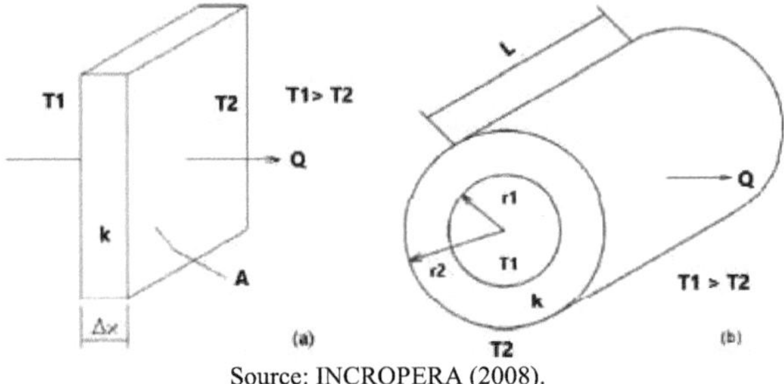

Source: INCROPERA (2008).

This defines the calculation of heat transfer by conduction as follows for flat plates and cylinders, respectively.

$$Q=-kA\frac{\Delta T}{\Delta x} \quad (2.2)$$

$$Q=2\pi kL\frac{\Delta T}{\ln\frac{r2}{r1}} \quad (2.3)$$

Where:

Q: Heat flux [W]

k: thermal conductivity [W/m.k]

A: area normal to the heat flow[m2]
T: temperature difference [K]
x: plate thickness [m]
r_1: inner radius of the cylinder [m]
r_2: outer radius of the cylinder [m]
L: length of cylinder [m]

2.2.2 Convective Heat Transfer

According to INCROPERA (2014) convection refers to the transfer of heat that occurs between a surface and a moving fluid when they are at different temperatures. This mode of heat transfer encompasses two mechanisms, namely the transfer of energy due to random molecular movement (diffusion) and the transfer of energy through the global, or macroscopic, movement of the fluid (INCROPERA. 2014).

Convective heat transfer is defined as the transfer of energy from the interior of a fluid, caused by the combined effects of conduction and global flow, to the macroscopic fluid. In this way, a convective heat transfer coefficient is presented, used to solve problems, based on the following equation:

$$Q=hA\Delta t \qquad (2.4)$$

Where:
Q: Heat flux [W]
A: area normal to the heat flow[m^2]
Δt: temperature difference [K]
h: convection coefficient [$W/m^2.K$]

2.2.3 Heat Transfer by Radiation

Radiation is defined as the energy emitted by matter in the form of photons (electromagnetic waves) as a result of changes in the electronic configurations of atoms or molecules. According to Cengel (2013), the transfer of heat by radiation differs from the others mentioned in that it does not require the presence of an intervening medium.

According to (INCROPERA, 2008), all surfaces with a different temperature emit energy in the form of electromagnetic waves. Therefore, in the absence of an enclosed medium, there is net heat transfer by radiation between two surfaces at unequal temperatures.

Therefore, the heat exchanged between two surfaces is calculated using the following formula:

$$E = \sigma T^4$$

(2.5)

$$Q = \varepsilon A \sigma (Ti^4 - Tf^4)$$

(2.6)

Where:
Q: Heat flux [W]
E: emissivity
Ti: surface temperature [K]
Tf: neighbouring surface temperature [K]
A: surface area [m]^2

2.2.4 Global Heat Transfer Coefficient

As an important parameter, the overall heat transfer coefficient (U) is used to calculate the thermal load. Table 3.3 of the book Instalagoes de Ar condicionado, CREDER (2004), sets out the values for the overall heat transfer coefficient applicable to certain types of walls. However, in the case of unconventional structures, characterised by the use of multiple different materials, the respective resistances inherent to each component are adopted for a more accurate calculation. These resistances, equivalent to the reciprocals of conductivity and conductance, are combined using a procedure analogous to the association of resistances in series observed in electrical circuits.

2.2.5 Psychometrics

It is now widely recognised that psychometrics is the science that investigates the influence of the characteristics of humid air (made up of a combination of dry air and water vapour) and the associated processes (such as drying, humidification, cooling and heating) on the change in temperature or the amount of water vapour present in the mixture.

The fundamental principles of psychometrics are directly applied in various contexts, including the calculation of the thermal load, the design of air conditioning systems, dehumidification and cooling coils, cooling towers and evaporative condensers (JESUS, 2002).

When developing projects, especially those related to air conditioning, understanding certain properties is crucial. These properties are known as psychrometric properties (JESUS, 2002).

When calculating the thermal load, fundamental psychometric parameters are used to define the thermal comfort range for human beings. These parameters include:

- Dry air: The composition of all atmospheric gases except water vapour;
- Saturated air and unsaturated air: Saturated air consists of the mixture

of dry air with saturated water vapour, while unsaturated air is the combination of dry air and superheated water vapour;

- Relative humidity: This represents the ratio between the saturation pressure corresponding to the temperature of the mixture and the partial pressure of the water vapour present in the mixture;
- Absolute humidity: Defined as the ratio between the mass of vapour and the mass of dry air, i.e. the total amount of water vapour in the atmosphere;
- Dry bulb temperature: The temperature indicated by an ordinary thermometer, without taking into account the influence of radiation;
- Wet-bulb temperature: The temperature recorded by a thermometer whose bulb is in contact with a wet wick. The reading at this point is called the wet bulb temperature of the air;
- Dew point: The temperature at which water vapour saturates and condenses.
- Specific volume of humid air: Calculated as the ratio between the volume of the mixture and the mass of dry air.

These parameters provide the basis for creating the so-called Psychrometric Chart, an essential tool for determining the appropriate thermal comfort range for air conditioning in a given room.

2.2.5.1 Psychrometric chart

The psychrometric chart is a visual representation of the parameters discussed above and plays an extremely important role in visualising the thermal comfort range for air conditioning. As a result, it enables the precise definition of the air conditioning cycle.

2.3 THERMAL COMFORT

According to Fanger (1970), thermal comfort is a condition of the mind that expresses satisfaction with the thermal environment. According to studies carried out by Silva (2015), thermal comfort is related to man's desire to feel good. Figure 4 shows the factors that influence human thermal comfort.

Figure 3 - Thermal comfort factors.

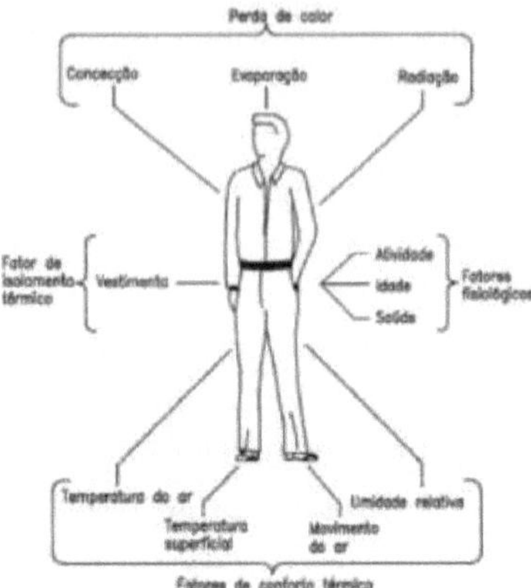

Source: Pizzeti (1970).

In order to relate thermal comfort and its influence on work performance, Frota and Schiffer (2001) state that the American Ventilation Commission carried out studies and research that confirmed the following results:

- For physical labour, an increase in the ambient temperature of 20°C to 24°C decreases yield by 15 per cent;
- 30°C ambient temperature with relative humidity of 80%, the yield drops by 28 per cent.

Thermal comfort is therefore a variable to be analysed when you want to consider improving working conditions and increasing productivity.

According to Frota and Schiffer (2001), humans need to release enough heat for their internal temperature to remain at an average of 37°C. The authors therefore associate this capacity with thermal comfort. "When heat exchanges between the human body and the environment take place without greater effort, the individual's sensation is one of thermal comfort and their work capacity, from this point of view, is maximum." (FROTA and SCHIFFER, 2001, p.15).

Therefore, if the individual feels hot or cold, it is because there is a thermal dysregulation, causing an additional effort to return to the initial condition, resulting in a drop in work performance.

2.4 ROOM AIR CONDITIONING

According to CREDER (2004), air conservation is an air treatment method whose purpose is to regulate air characteristics such as temperature and relative humidity in a given environment.

Air conservation, according to NBR 16401, requires the management of the quantities described below, corresponding to the desired state within the environment, which must be maintained throughout the entire period of operation of the air conditioning system:

- Relative humidity;

- Air movement;
- Air temperature;
- Air purity level;
- Air renewal percentage.

In addition, there are other concepts for air conditioning that are key factors in system design:

- Temperature: Refers to the equality of temperature between two bodies and/or environments;
- Pressure: Ratio of normal force to area. The pressure at all points of a system in equilibrium is the same in any direction;
- Volume: This can be presented as the inverse of the specific mass;
- Heat: Thermal energy that flows exclusively when there is a temperature differential;
- Latent heat: The amount of heat that is removed from or added to an environment, changing its state without affecting its temperature (CREDER, 2004);
- Sensible heat: The amount of heat in a room that must be removed or added due to the temperature difference between the outside and inside, in order to promote the desired comfort conditions (CREDER, 2004);
- Air Dry Bulb Temperature: Temperature related to the air-vapour compound. It has the same value for both elements of the compound;
- Humid Bulb Air Temperature: Temperature less than

The dry bulb temperature is obtained using a thermometer.

2.5 THERMAL CHARGE

Considering the thermal load is essential when designing an air conditioning system. To do this, some related specifications must be adopted in projects and these are indicated in ABNT NBR 16401. It is calculated taking into account environmental factors such as external and internal temperatures, solar radiation, internal heat gain and room humidity.

Also according to ABNT NBR 16401, the *ASHRAE Handbook is* suggested, which cites the RTS (*Radiant Time Series*) model as a reference for calculating the thermal load. This method has led to the creation of computer software for determining thermal loads.

For the correct sizing of the air conditioning system, it is important to analyse the ideal conditions for human comfort and to calculate the thermal load. In this study, the CYPETHERM LOADS software will be used to predefine the factors that affect the loads for the particular region in which the establishment is located.

It is essential to determine various characteristics of the environment in order to carry out the appropriate calculation. The characteristics to be defined include:

- Location of the environment: This involves its geographical position, as well as exposure to the sun and shaded areas;
- Room category: Identify whether it is residential, commercial, office, hotel or other;
- Physical dimensions of the room: Consider the height, length and width of the space;
- Building Materials: Evaluate the materials used in construction, such as the type of brick, glass, windows, roof or slab, and how these materials affect heat exchange;
- External conditions: Take into account elements such as the façade, colours and shades present in the external environment;
- People: Determine the number of people in the room;
- Lighting: Specify the type and quantity of lighting present in the space;
- Equipment: Account for the quantity and power of equipment installed in the environment;
- Doors: Consider the location and number of doors in the espago;
- Temperature: Collect information on dry bulb temperature, wet bulb temperature and relative humidity in the room;
- Heat Transmission: Evaluate the heat transmission caused by solar exposure, both indoors and on the external wall and roof;
- Insulation: Studying how heat is transmitted through internal glazing, walls and floors;
- Lighting: Consider the impact of lighting on the environment and how it affects heat;
- Occupant load: Analyse the thermal load generated by the people present in the space.

Analysing and defining these characteristics are crucial steps
20 to make accurate and effective calculations.

2.5.1 Dry Bulb Temperature, Humidity and Relative Humidity

Once all the characteristics of the thermal zone have been surveyed, the respective dry bulb and wet bulb temperatures of the external and internal zones must be defined. With regard to external conditions, table A.6 (annex) of the NBR 16401 standard should be used, which, based on the chosen location, defines the variables of TBS, TBU, atmospheric pressure and absolute humidity. However, the internal conditions are determined by the NBR 164012 standard. The absolute indoor humidity is defined using the expression:

$$W = 0{,}622 \frac{Ps}{Pt - Ps} \qquad (2.7)$$

Where:
Ps: partial pressure of water vapour
Pt: atmospheric pressure
2.5.2 Thermal load caused by insolation
Solar energy is, in most cases, responsible for the largest share of the total thermal load in air conditioning calculations, usually as convection and radiation.
2.5.2.1 Heat transmission from the sun through opaque surfaces
The roof, slabs and walls of the enclosure transmit solar energy to the interior by means of conduction and convection, as shown in the following formula

$$Q = AU[(Te - Ti) + \Delta t] \qquad (2.8)$$

Where:
A: area of the opaque surface [m^2]
U: global heat transmission coefficient [$W/m^2 °C$]
Te: outside temperature [°C]
Ti: internal temperature [°C]
At: increase in temperature due to insolation [°C]
2.5.2.2 Heat transmission from the sun through transparent surfaces
In the bibliography Instalagoes de Ar condicionado, CREDER (2016), you can find the solar factor values obtained through experiments, given in BTU/h or W/m^2 , assuming that the window has no protection. If it has awnings or blinds, multiply the values found by the respective reduction coefficients:

- External blinds or awnings: 0,18 - 0,2
- Reflectors or internal blinds: 0.58 - 0.67
- White interior curtains (dull): 0.43 - 0.49

Based on these parameters, we can determine the transmission of heat from the sun, according to the following equation:

$$q = A fator \qquad (2.9)$$

Where:
A: transparent surface area [m^2] factor: solar factor [BTU/h or W/m]^2
2.5.3 Thermal Load Due to People
According to NBR 16401, the thermal load due to people is given by the sensible and latent heat released by people in an enclosure, for a given dry bulb temperature of the internal conditions. The standard also gives values for the sensible and latent heat released by people depending on the temperature and the activity carried out at a dry bulb temperature of 24°C. However, it can be

calculated for any dry bulb temperature, as shown in equation 2.10.

$$Qpessoas = n(qs + ql) \qquad (2.10)$$

Where:
n: number of people in the thermal zone
qs: sensible heat [W]
ql: latent heat [W]

2.5.4 Thermal Load Due to Equipment

According to NBR 16401, the thermal load due to the equipment is obtained through heat dissipation values (office equipment, motors and other heat humidity sources) and dissipated power.

2.5.5 Thermal Load by Air Renewal

In order to determine the air renewal rate, NBR 16401-3 is used as a reference, labelled indoor air quality.

This part of the standard stipulates the minimum flow of outdoor air of acceptable quality to be supplied by the system in order to renew the indoor air and keep the concentration of pollutants at low levels. This rate is determined using the ASHRAE methodology, and it is first necessary to calculate the air flow required for renewal, given by equation 2.11:

$$Vef = Pz * Fp + Az * Fa \qquad (2.11)$$

Where:
Vef: effective outside air flow [L/s]
Fp: flow per person [L/s person]
Fa: usable area occupied by the person[m]^2
Pz: maximum number of people in the ventilation zone

2.6 RADIANT TIME SERIES METHOD (RTSM)

With the aim of using the most modern tools in the field, CYPETHERM LOADS software operates in a BIM flow that meets ASHRAE recommendations.

According to Mcquiston (2005), the RTSM is based on the Heat Balance Method and does not apply iterative techniques. As such, it uses mathematical series in order to physically partition environments (SPITLER, 2020).

In this way, heat gains are calculated individually for each type of boundary using 24 response factors, analysing each hour of the day. These factors provide a time series solution for the one-dimensional and transient conduction heat transfer problem (SPITLER, 2020).

$$q\theta = A\Sigma YPj(te, \theta - j\delta - trc)23j = 0 \qquad (2.12)$$

Where:

$q\theta$: conductive heat transfer to surface [W]
A: surface area [m]^2
YPj: response factor for relative surface area
$\theta - j\delta$: SUN - AIR temperature [°C]
trc: air temperature of the room, constant [°C]

2.7 EQUIPMENT FLOW

In line with Soares (2007), volumetric flow corresponds to the volume of fluid passing through the area of a given section per unit of time, conventionally expressed by the unit (m^3 /s).

Figure 4 - Representation of pipeline flow.

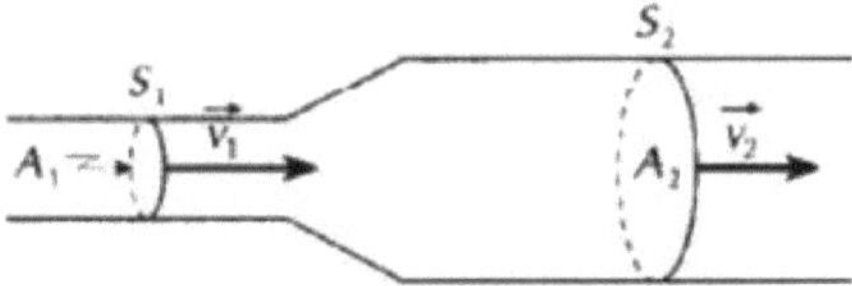

Source: Principles of Fluid Mechanics.

With regard to heat flow, this will depend on the difference between the internal and external temperatures, the properties of the fluid and the thermal load. Thus, by obtaining the Thermal Load and the temperature outside the room, you can calculate the air conditioning flow required to remove enough heat to maintain the desired temperature.

To obtain this quantity, equation 2.13 is used (FOX, 2014).

$$Q = \frac{X}{c * \Delta T * \rho} \quad (2.13)$$

Where:
Q: flow rate (m^3 /s)
X: total thermal load (W)
c: specific heat of the fluid (J/kg*°C)
ΔT: external-internal temperature difference (°C)
ρ: specific mass of the fluid (kg/ m^3)

2.8 MODELLING IN A BIM ENVIRONMENT

The concept of BIM (*Business Intelligence Modelling)* brings together the idea of constructing a virtual building before actually building it, i.e. modelling construction information. In this way, this prior modelling allows for a process of generating and sharing construction information in a way that offers benefits at all stages of the project's life cycle (PEREIRA, 2017).

Guaranteeing the life cycle and evaluating it is fundamental in design practice, as it allows for greater guidance in decision-making in the project and in the search for advanced solutions. The BIM environment is used in the professional world mainly for these purposes.

Masotti (2014) states that BIM has several layers of information, its dimensions. Depending on the operator's needs, a model in this software can be generated from 4D onwards.

3 METHODOLOGY

The project methodology uses software and tools that provide concise results. In order to guide the case study, the structure of the work was built. The case study was conceived and composed of 4 stages (problem analysis, problem modelling, calculations in the shop environment, analysis of the values obtained), as described in the flowchart below:

Figure 5 - Project flowchart.

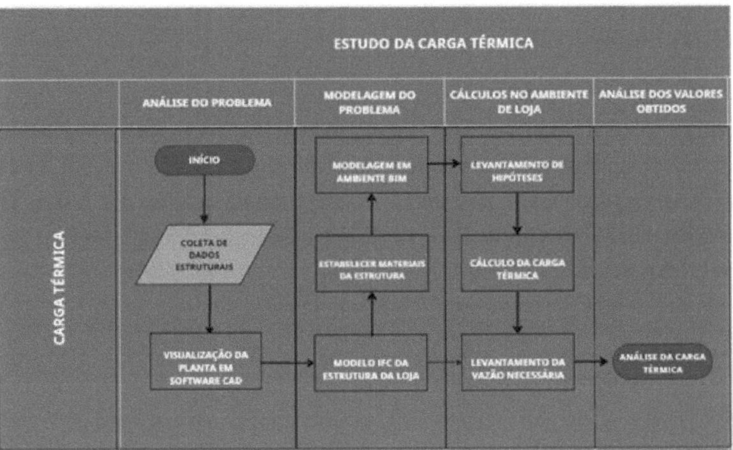

Source: Author.

3.1 BIM ENVIRONMENT

The search for the most up-to-date practices has led to the proposal of a collaborative work cycle project. The use of computer tools such as Building Intelligence Modelling (BIM) represents a new generation of software focused on the project life cycle (MARCOS, 2009). The main directions seek interoperability and the reuse of information with the aim of continuous improvement in the project development process. Thus, the BIM flow is the best solution, as it allows for higher quality information that will be integrated throughout the cycle (EASTMAN, 2014).

In the context of this study, BIM was used to create a complete digital model of the motorbike shop in question. This consisted of a detailed representation of structural elements such as floors, walls, ceiling and ventilation system, as well

as internal information on
layout, shelving, counters and displays. Once this information has been entered, the technical analysis of the problem begins.

The great advantage of using BIM was the ability to simulate internal conditions and scenarios that used to be carried out, as a rule, by dynamic spreadsheets, resulting in an increase in project simulation time. In addition, the computational tool allows the integration of external climate data, which made it possible to evaluate the impact of meteorological conditions on the shop's internal temperature.

3.2 ANALYSING THE PROBLEM

The problem analysis phase of this work involves a detailed study of the problem in question, which is the thermal load in a motorbike shop. This context is fundamental to identifying the critical points that influence the shop's internal temperature and thus guide the subsequent stages of the project.

To begin with, structural information about the building was collected. To do this, the engineering team of the company responsible for the shop provided data such as specifications for building materials, area dimensions, type of roof and ventilation system. This data was essential for understanding the starting conditions and creating the BIM model.

In addition, the building plan was developed using CAD (*Computer-Aided Design*) software. This software makes it possible to represent the architectural features of the environment, as it takes into account all the details provided by the engineering team. This includes the layout of key elements in the regulation of the thermal load.

Therefore, the problem analysis also considered the Coordination's specifications, ensuring that all requirements and particularities were incorporated into the project. This includes, for example, accessibility requirements for clients.

3.3 MODELLING THE PROBLEM

Once the previous stage had been completed, the next step involved modelling the building. In this context, modelling refers to the creation of a detailed three-dimensional model of the motorbike shop using BIM.

The *IFC Builder* software was the tool responsible for this modelling process. It enabled the creation of a complete digital model, where every element of the building is accurately represented. This includes walls, floors, ceilings, pillars and all other components relevant to the shop's structure.

And this three-dimensional representation is crucial because it enabled a more in-depth analysis of the internal thermal conditions. With the BIM model, it was possible to visualise how the air circulated inside the shop, identify hot and cold spots, and also assess the effectiveness of the ventilation systems.

In addition, BIM has also incorporated information about building materials. This means that the thermal conductivity of the materials present, their insulation capacity and other factors related to the thermal load can be taken into account. These details are essential for understanding how the structure of the shop affects the thermal behaviour of the environment. After creating the 3D model, the structural plan is integrated into the *Cypetherm* software.

Loads. This specialised tool is designed to evaluate the structural properties of buildings and analyse the elements that can influence the building's thermal load. With the integration of the model, *Cypetherm Loads* became the tool adopted to calculate the thermal load of the motorbike shop.

Thus, the modelling of the problem was not limited to the visual representation of the shop, but also incorporated essential technical data for the analysis of the thermal load. This comprehensive approach was fundamental to understanding how the structure of the shop and its components affect the internal thermal environment.

3.4 CALCULATIONS IN THE SHOP ENVIRONMENT

The main aim of the calculations carried out in the shop environment was to determine the thermal load, i.e. the amount of heat entering and leaving the space over a given period of time. This calculation is essential for assessing whether the existing air conditioning system is capable of maintaining the internal temperature within comfortable limits for the occupants.

In addition, these calculations also made it possible to identify critical areas where the thermal load could be excessive, such as places with a high concentration of equipment or direct exposure to sunlight. With this information in hand, it is possible to develop strategies to optimise the energy efficiency and thermal comfort of the shop, whether by adjusting the air conditioning systems, reorganising the layout or improving the thermal insulation.

Therefore, the calculation phase in the shop environment served to quantify and understand the thermal load of the motorbike shop. Detailed analysis of all the internal and external factors that influence temperature is fundamental to making informed decisions about efficient thermal load management, with a view to customer and employee comfort.

4 RESULTS

4.1 Case Study

In order to prove the above methodology, a case study with the following specifications will be adopted:

A shop in the city of Recife - PE with a medium flow of people and fully functioning electronic equipment. To this end, it is necessary to make it feasible, by obtaining the thermal load, to size an air conditioning system for the shop, which currently only has medium-sized fans. Thus, taking into account the thermal comfort of employees and customers, the assertiveness of the thermal load study guarantees the implementation of an effective system for the shop to offer a better quality service.

4.1.1 Climate data

Weather Data was used to obtain climate data.

View 6.0 in which we inserted the highlighted parameters (Figure 6) referring to the city where the shop is located (Recife - PE).

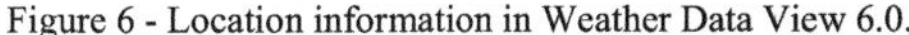
Figure 6 - Location information in Weather Data View 6.0.

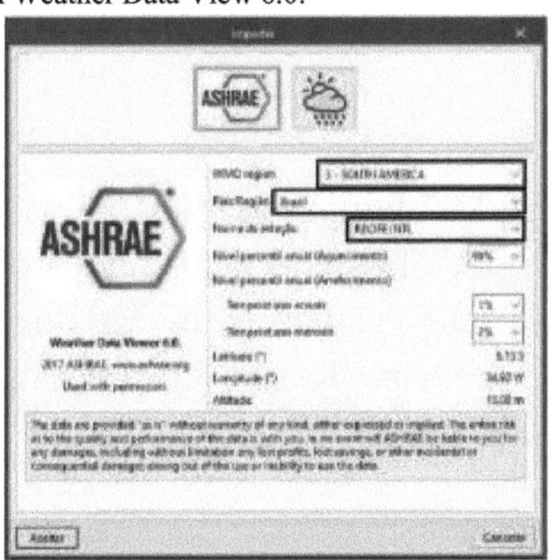

Source: Author.

As it is a commercial environment, its operation is uninterrupted throughout the year.

throughout the year. For this reason, the average temperature values per month adopted were obtained from ASHRAE's *Clear-Sky Solar Radiation* model, as shown in Figure 7. This fulfils the needs of this study.

Figure 7 - Average temperature data throughout the year.

	Dry design temperature ГC)	Coincident humid temperature (°C)	Daily dry temperature oscillation (-C)	Daily oscillation of the humidity temperature	Direct irradiation	Diffuse radiation
JANUARY	33,2	26,4	5,4	2.6	0.394	2.503
FEBRUARY	33,8	26,9	5,6	2.7	0.398	2.497
MAR(JO	33,9	27,2	5,9	2.8	0.397	2.507
APRIL	32.9	26,8	6,1	2.9	0.393	2.538
MAY	32,1	26,4	6,0	3	0.386	2.559
JUNE	30,9	25,6	5,8	3.1	0.39	2.534
JULY	30,2	24,9	6,0	3.3	0.391	2.517
AUGUST	30,2	24,7	5,9	3.2	0.395	2.471
SEPTEMBER	31,0	25,1	5,4	2.9	0.425	2.361
OCTOBER	32,2	25,6	5,2	2.8	0.421	2.377
NOVEMBER	32,9	26,0	5,2	2.8	0.399	2.451
DECEMBER	33,2	26,2	5,3	2.8	0.404	2.455

Source: Author.

4.1.2 Internal heat sources

4.1.2.1 Heat gain from lighting

With regard to shop lighting, the data was obtained directly from the Cypetherm Loads software, which has standardised data in its database according to the zone of the building under study and the category of luminaire used in the premises, as shown in Figure 8.

Figure 8 - Input information for lighting power.

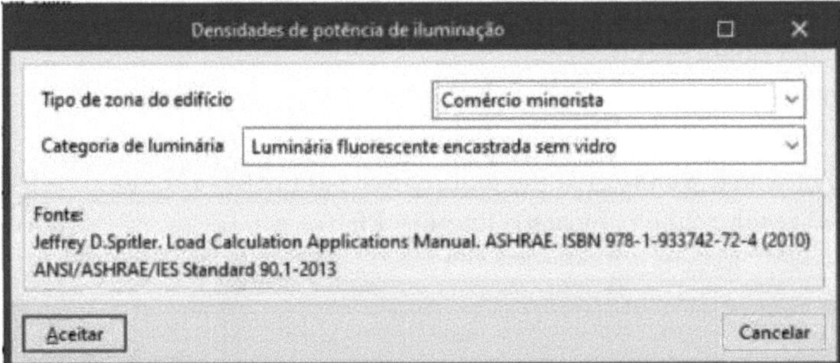

Source: Author.

By selecting the type of building zone (retail) and category of luminaire (recessed fluorescent without glass) used in the shop in this study, you can obtain the results for heat gain related to the lighting of the premises. These values are found in the Cypetherm loads software database and use ASHRAE data as a library, as shown in Figure 9.

Figure 9 - Lighting power information.

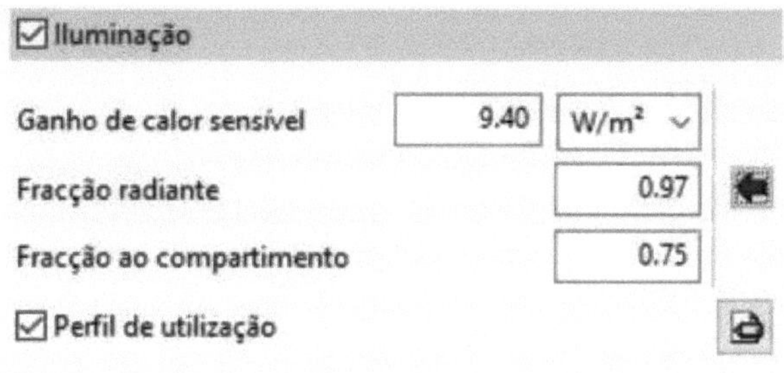

Source: Author.

4.1.2.2 Heat gain per occupancy

The shop in this study is located in the city centre and offers face-to-face services. To analyse the case study, heat gain per occupancy was taken into account, with input data selected according to Figure 10:

Figure 10 - Input information for internal heat gain by occupancy.

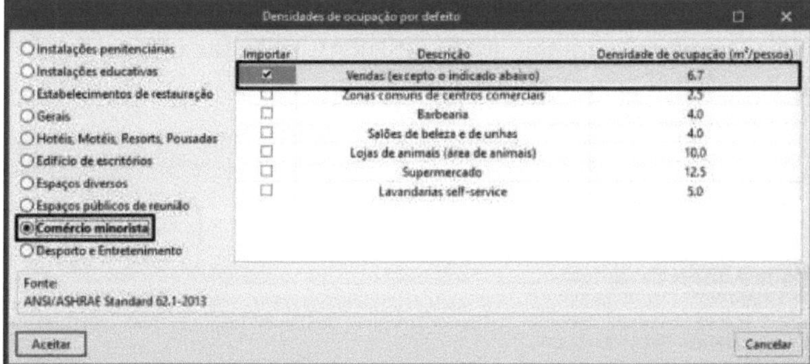

Source: Author.

Once the type of establishment has been chosen, it is necessary to enter the level of activity (shopping centre; shop) that takes place within the commercial environment. In addition, the program offers the possibility of setting different percentages for men and women according to the reality of the shop being analysed. However, as this is a commercial environment, there is no way of predicting the degree of occupation separately. Therefore, equal percentage values will be adopted, as shown in Figure 11:

- Percentage of men, women: 50%

Figure 11 - Information on the level of activity in the shop.

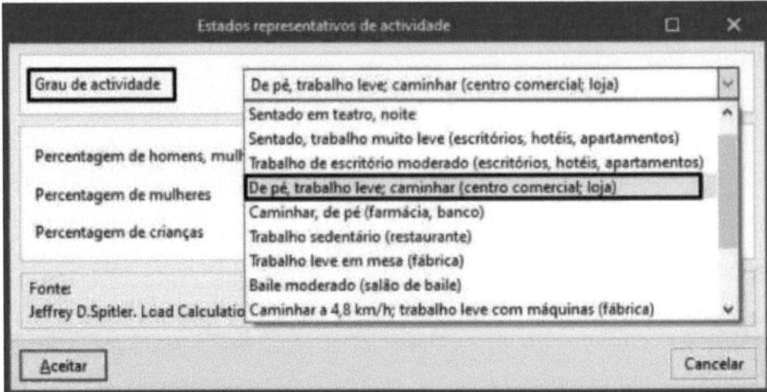

Source: Author.

4.1.2.3 Heat Gain from Internal Equipment

With regard to equipment that can provide internal heat gain, the shop has a desktop computer, which is responsible for cashier operations (selling and removing products) and a notebook computer, which is used for administrative activities. Therefore, according to the ASHRAE table, the office's thermal load density is classified in the light category, see Figure 12:

Figure 12 - Information on internal heat gain by equipment.

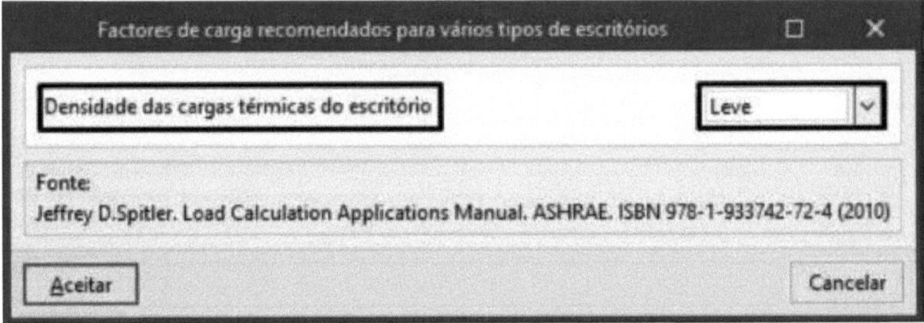

Source: Author

With this input information, the software provides the ASHRAE database with the values for the type of thermal load density of the selected office, as shown in Figure 13. In a meeting with the company's staff, the conclusion was reached that the density of thermal loads caused by equipment is light, given that the electronic media used are only a desktop computer (used at the counter to record products sold and check nails) and a notebook (the shop manager's work medium).

Figure 13 - Input information for internal heat gain by equipment.

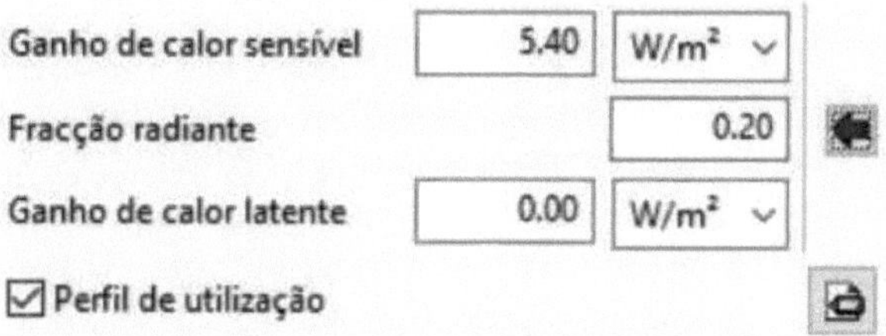

Source: Author

4.1.3 Store environment

The enclosure was built according to the characteristics requested by the company's coordinators. The CAD plan and the composition of the materials to be used were supplied by the team of engineers responsible for the project. It should be emphasised that the space built has an upper floor, where all the product stock is located. The structure of the shop is shown in Figure 14, with its structural data below:

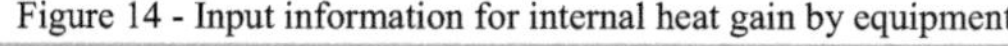
Figure 14 - Input information for internal heat gain by equipment.

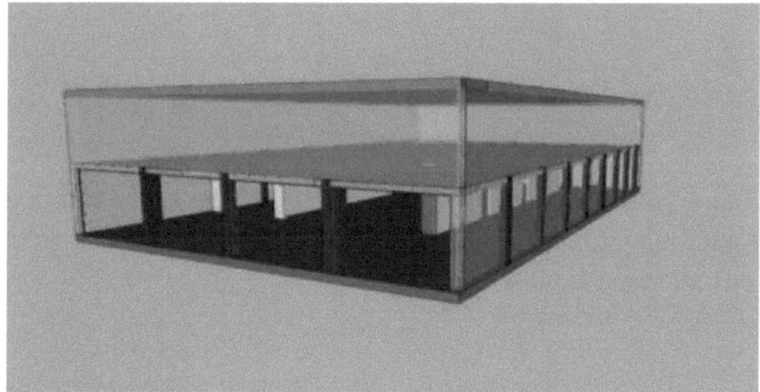

Fonte: Autor.

- Venue volume: 1324.9 *metres*^{3}
- Total area: 473.2 m^2
- External wall width: 0.3 m

Internal wall width: 0.15 m

The views of the shop environment are shown below. For identification purposes, they are represented by acronyms:

- **WALL-A**: External wall Opening
- **F-WALL**: Outer wall Bottom
- **D-WALL**: Right Side External Wall
- **WALL-E**: External Left Side Wall

The sun's incidence on the enclosure reaches perpendicularly to the External Wall Opening.

Figure 15 - Front view of the enclosure disregarding doors.

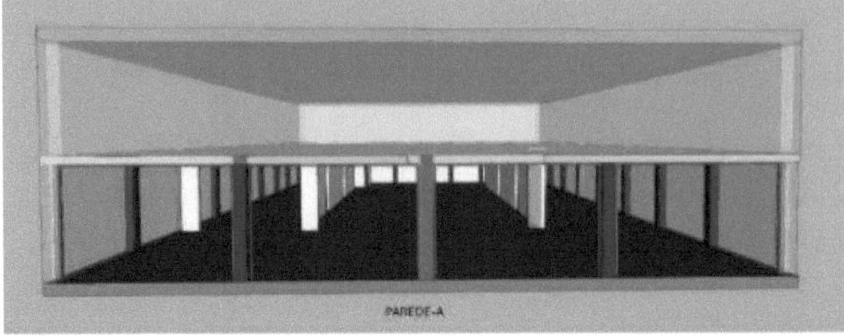

Source: Author.

Figure 16 - Rear view of the enclosure.

Source: Author.

Figures 15 and 16 show, respectively, the front of the shop, where customers access the main environment, and the back of the shop, a space used only for employees to enter and leave.

Figure 17 - Right side view of the enclosure.

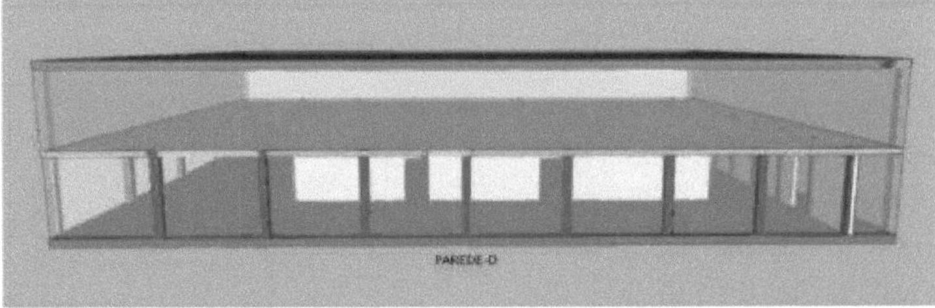

Source: Author.

Figure 18 - Left side view of the enclosure.

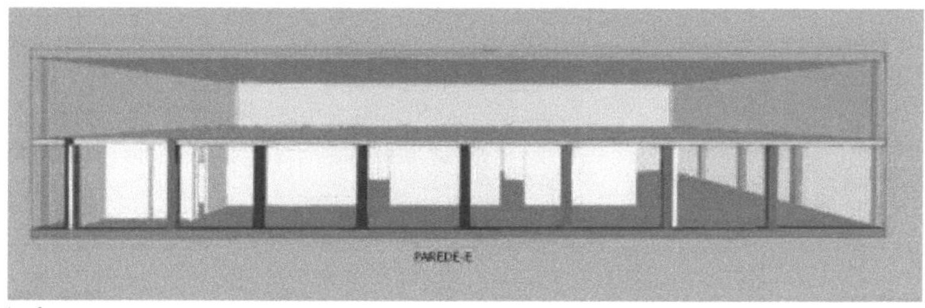

Source: Author

The external walls, which are perpendicular to the ground, have a structural composition as shown in Figure 19 and their thermal characterisation was obtained from the ISO-6946 standard (2017).

Figure 19 - Left side view of the enclosure.

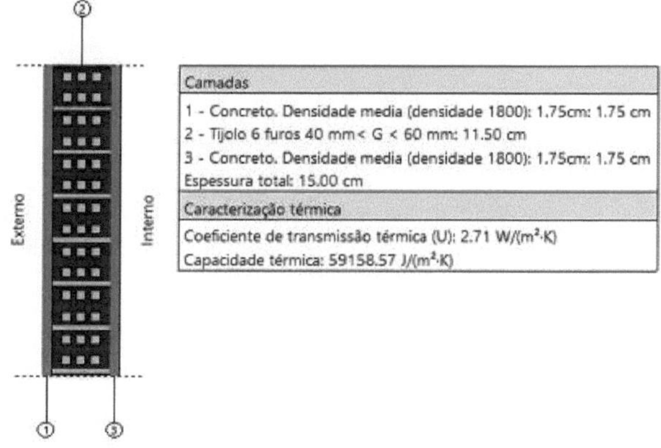

Source: Author.

WALL-I: Lower External Wall (Floor)

WALL-P1: Internal wall between floors (slab)

The Lower External Wall (WALL-I) and the Internal Wall Between Floors (WALL-P1) are structurally composed as shown in Figures 20 and 21:

Figure 20 - Structural composition of the lower external wall (floor).

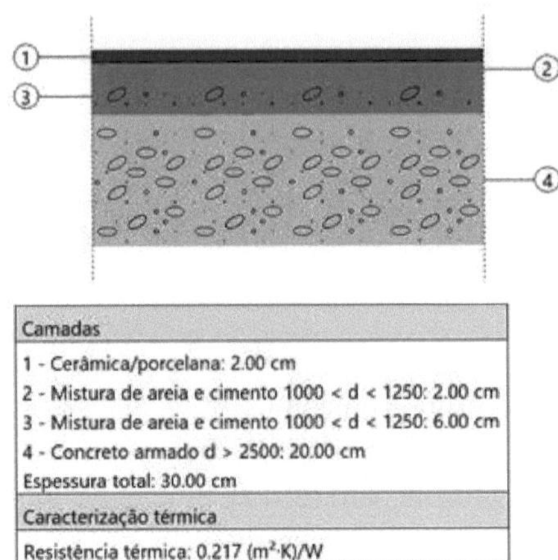

Source: Author.

Figure 21 - Structural composition of the internal wall between floors (slab).

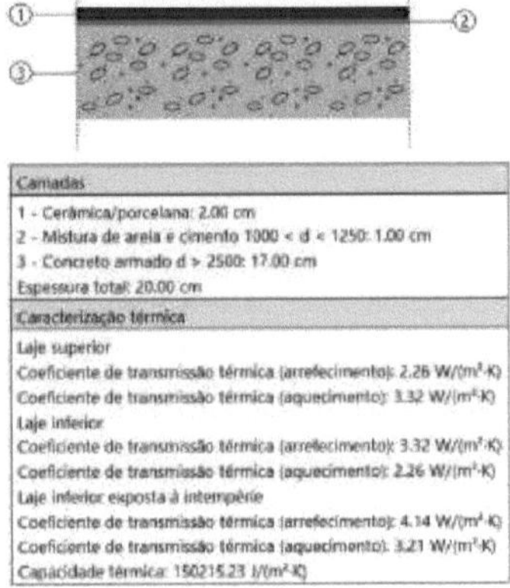

Source: Author.

The thermal characterisations shown in figure 22 were obtained using the ISO-6946, ISO 10077, ISO 10456 and ISO 13370 standards:

LAYER COMPONENTS	Specific heat capacity Cp	RT Thermal Resistor	Density p (kg/m3)	Layer E thickness (cm)	Thermal Conductivity A(W/(m*K))
Mixture of sand and cement 1OOO < d < 1250	1000.0	0,04	1125.0	2.0	0,55
Reinforced concrete d > 2500	1000.0	0,04	1800.0	10el7	2,5
6-hole brick 40 mm < G < 60 mm	1000.0	0,17	1140.0	11.5	0,68
Ceramics/Porcelain	1000.0	0,01	2000.0	2.0	2.0

Source: Author

4.2 PROPOSALS FOR USING THE ENVIRONMENT

The proposals for using the environment were adopted as follows:

- **Heat gain per occupancy:**

Occupancy density: 6.7 m^2 /person

Radiant Fragao: 58%

Latent heat: 55 W/person

- **Heat gain by internal equipment**

Sensitive heat: 150 W

Radiant Fragao: 50%

Latent heat: 150 W

- **Heat gain from lighting :**

Sensible heat: 13.6 W/m^2

Radiant Fragao: 58%

- **Daily profile by occupation (01)**

07:00 to 12:00 (gradual increase in the number of customers - shift shift)

12:00 to 13:00 (time with the highest number of customers in the shop - peak time in the shopping centre)

13:00 to 18:00 (decrease in the flow of customers in the shop)

Figure 23 shows the proposals for heat gain in the environment and the daily use profile by occupancy.

Figure 23 - Daily profile by occupation.

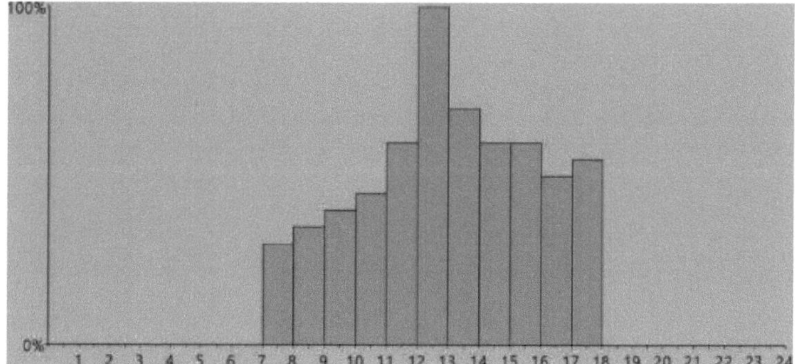

Source: Author.

4.3 THERMAL LOAD VALUES

4.3.1 Results by border

1. External walls:

Constant of the Radiant Time Series Method throughout the day

Figure 24 - RTSM external wall constants.

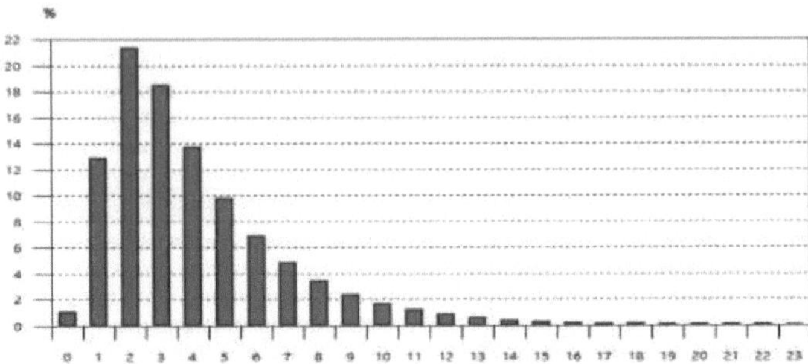

Source: Author.

- Variation in total solar irradiance (yellow), external temperature (green) and sun-air temperature

Figure 25 - Temperature variation and solar irradiation on external walls.

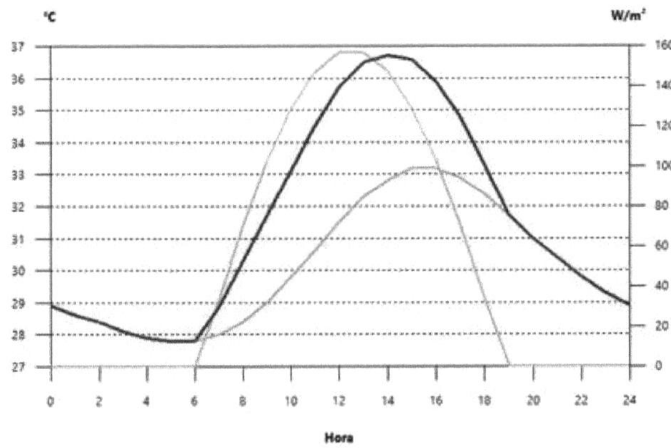

Source: Author.

- Variation of thermal load (brown) and outside temperature.

Figure 26 - Variation in the thermal load on external walls.

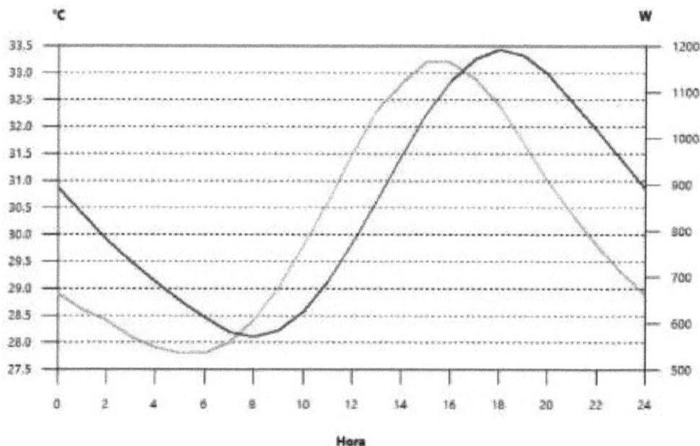

Source: Author.

2. Slab (Floor between floors):

Variation of thermal load (brown) and external temperature (green)

Figure 27 - Variation of the thermal load on the slab (floor between floors).

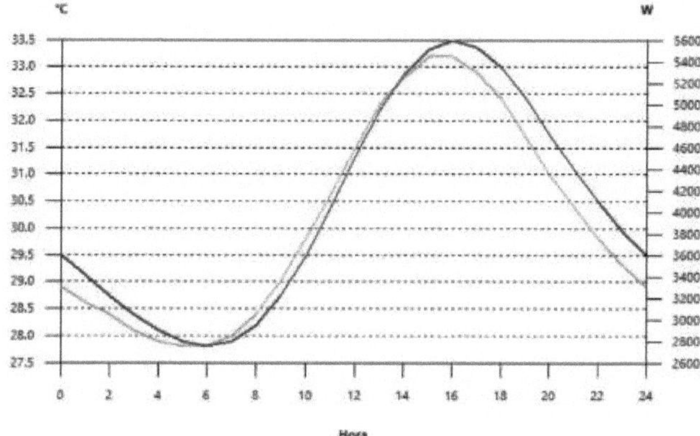

Source: Author.

3. Roof

Constants of the Radiant Time Series Method for each hour of the day:

Figure 28 - RTSM constants on the roof.

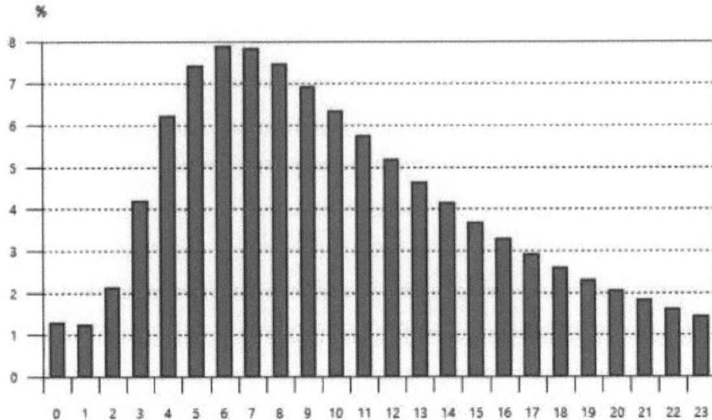

Source: Author.

- Variation in total solar irradiance (yellow), external temperature (green) and sun-air temperature (brown):

Figure 29 - Temperature variation and solar irradiation on the roof.

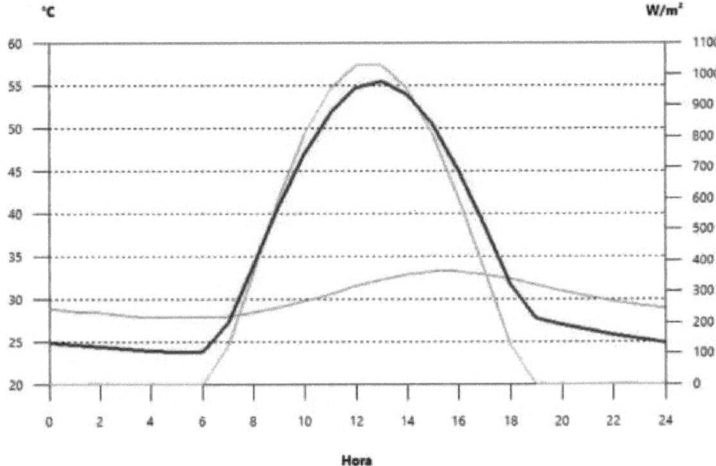

Source: Author

Variation in thermal load (brown) and external temperature (green):

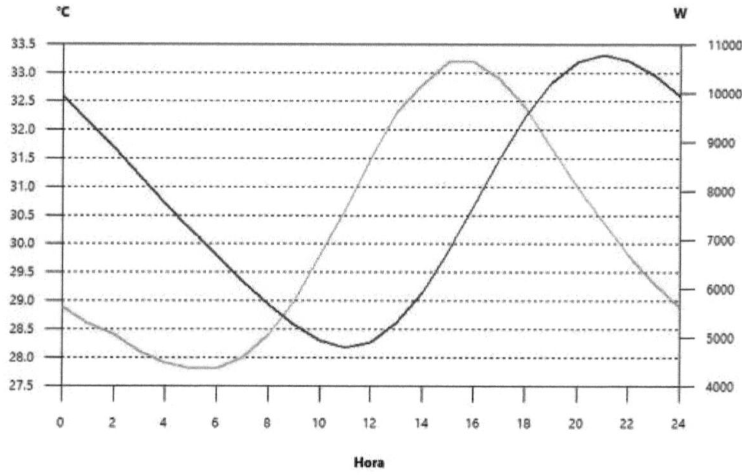

Figure 30 - Variation in thermal load on the roof.

Source: Author.

4. External walls (glass door) - Variation in thermal load through glass doors

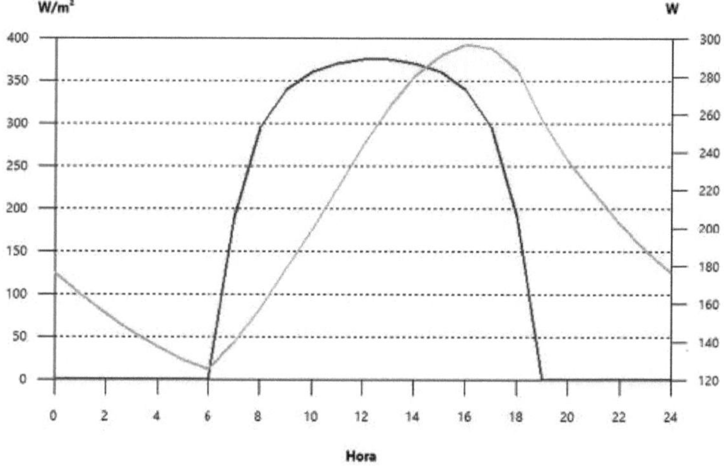

Figure 31 - Variation in thermal load on glass doors.

Source: Author.

4.3.2 Calculating the thermal load due to air renewal

Once the data had been collected using the software, the necessary calculations were made to find the air flow needed to renew the indoor air and keep the concentration of pollutants at an acceptable level. Excel software was used to find these values, see Figures 32 and 33:

Figure 32 - Calculation of the air flow required for renovation (shop environment).

Thermal Load by Air Renewal

Reference ABNT NBR 16401-3

I) Calculation of the Air Flow Required for Renovation (Shop Environment)

Carga Térmica por Renovação de Ar		
Referência ABNT NBR 16401-3		
I) Cálculo da Vazão de Ar Necessária para Renovação (Ambiente Loja)		
Dados	**Valor**	**Definição**
Vef [L/s]	X	Vazão Eficaz de Ar Exterior
Fp [L/p*pessoa]	5,7	Vazão por pessoa
Fa [L/S *m^2]	0,9	Vazão por área útil ocupada
Pz	8	Número máximo de pessoas na zona de ventilação
Az [m^2]	220	Área útil ocupada pelas pessoas
Vef [L/s]	**243,6**	

Fonte: Autor.

Source: Author.

Figure 33 - Calculation of the Air Flow required for renovation (Store Stock).

Thermal Load by Air Renewal

Reference ABNT NBR 16401-3

I) Calculation of the Air Flow Required for Renovation (Store Stock)

Carga Térmica por Renovação de Ar		
Referência ABNT NBR 16401-3		
I) Cálculo da Vazão de Ar Necessária para Renovação (Estoque Loja)		
Dados	**Valor**	**Definição**
Vef [L/s]	X	Vazão Eficaz de Ar Exterior
Fp [L/p*pessoa]	5,7	Vazão por pessoa
Fa [L/S *m^2]	0,9	Vazão por área útil ocupada
Pz	4	Número máximo de pessoas na zona de ventilação
Az [m^2]	253,2	Área útil ocupada pelas pessoas
Vef [L/s]	**250,68**	

Fonte: Autor.

Source: Author.

To carry out the calculation, we used the table that deals with the minimum effective flow of outside air for ventilation, found in NBR 16401 part 3, as shown in figure 34:

Figure 34 - Table showing the minimum effective flow of outside air for ventilation.

Location	D people/ 100 m^2	**Level 1**		**Level 2**		**Level 3**	
		FP L/s'pess	F. L/s'm^2	FP L/s'pess	F. L/s'm^a	Fp L/s'pess	F. L/s'm^2
Retail trade							
High-end supermarket	8	3,8	0,3	4,8	0.4	5.7	0.5
Medium-sized supermarket	10	3.8	0.3	4,8	0,4	5.7	0.5
Popular supermarket	12	3.8	0.3	4.8	0.4	5.7	0.5
Shopping centre *mall*	40	3,8	0.3	4.8	0.4	5,7	0,5
Shops (except below)	15	3,8	0,6	4.8	0.8	5.7	0.9

Beauty skirt and/or barbershop[b]	25	10	0.6	12,5	0.8	15,0	0.9
Pets[b]	10	3,8	0.9	4.8	1.1	5.7	1.4
Self-service laundry	20	3.8	0.3	4.8	0.4	5.7	0.5

Source: Adapted (ABNT NBR 16401-3, 2008).

Next, after determining the air flow required for renewal, it is now necessary to determine the air flow to be supplied to the ventilation zone. This value was calculated and obtained as shown in Figures 35 and 36.

Figure 35 - Calculation of the Air Flow to be supplied to the Ventilation Zone (Store Environment).

Thermal Load by Air Renewal

: **Reference ABNT NBR 16401-3**

II) Calculation of the Air Flow to be supplied to the Ventilation Zone (Shop Environment)

Data	Value	Definite
Vz	X	Outside air flow to be supplied
Vef	243,6	Effective outside air flow
Ez	1	Air Distribution Efficiency
Value of Vz	243,6	

Source: Author.

Figure 36 - Calculation of the Air Flow to be supplied to the Ventilation Zone (StockStore).

Thermal Load by Air Renewal

Reference ABNT NBR 16401-3

II) Calculation of the Air Flow to be supplied to the Ventilation Zone (Store Stock)

Data	Value	Definition
Vz	X	Outside air flow to be supplied
Vef	253,2	Effective outside air flow
Ez	1	Air Distribution Efficiency
Value of Vz	253,2	

Source: Author.

After determining the flow rate, we find the sensible and latent heat
This gives the thermal load rate generated by air renewal,
as shown in figure 37:

Figure 37 - Thermal load rate generated by air renewal in the rooms.

Environment	CTrenovataosensivel IkWj	CTrenovfKaotertnrt IkWj	CT *ren ova^aoTotai* [kW]
Shop environment	3,08	7,73	10,81
Shop stock	3,20	8,04	11,24

Source: Author.

4.3.3 Total values

In addition, to carry out the analyses, *Data Weather View 6.0* calculates the maximum thermal load values from the day with the highest average temperatures and the highest solar irradiation rates, which in this study is represented by 21 March at 5pm, as shown in Figure 38:

Figure 38 - General Thermal Load values calculated for the hottest day of the year.

Summary of zone cooling loads: Zone 3

	External	Interns	Ventilation	Totals
	Solar conduction Inf. lat. Inf. sens. (m^2)(W)(W)(W) (W)	Lat. Sens. (W) (W)	Lat. Sens, (l/s)(W) (W)	Lat. Sens. Total Total (W) (W) (W/m^2) (W)
Maximum cooling load per compartment				
Environment Store Stock	473993720100 48819243000	4054 8809 4172 9072	554 16053 5471 570 16189 5287	20107 244199444525 20361 3360211153962

Maximum simultaneous cooling load for o set of rooms: 21 March at 5pm (17 apparent solar time)

Source: Author.

4.4 EQUIPMENT FLOW

Given the maximum thermal load over the months of the year and the other thermodynamic values of the fluid (Air), the flow rate needed to keep the room at the required ambient temperature is 37434.876 m^3 /h, as shown in Figure 39:

Figure 39 - Flow required for equipment.

Total Thermal Load (W)	98487
Specific heat of the fluid (J/(Kg*C^0))	1000
Maximum ambient temperature (°C)	25
Temperature Recife (°C)	33
Temperature variation (°C)	8.00
Specific mass of the fluid	1.18
Flow (m^3 /h)	37434.876

Source: Author.

4.5 OTHER HYPOTHESES

In order to address different construction hypotheses, other factors have been considered, which are described below:

Hypothesis 1 - building to the right of the shop

As the establishment is to be built in a shopping centre, the analysis of this hypothesis involves the construction of a building to the right of the shop, as shown in figure 31:

Figure 40 - Shop in a shopping centre.

Source: Author.

Considering the building next to the establishment, the cooling load results obtained are shown in Figure 41:

Figure 41 - Cooling loads with building next door.

	External				
	Solar conduction Inf. lat. Inf. sens. (m2) (w)(W)(W) (W)				
Maximum cooling load per compartment					
Environment Shop	473	9344	201	0	0
Stock Shop	488	18453	0	0	0

Source: Author.

Thus, comparing the initial proposition with the hypothesis above, it can be seen that the external loads are very close, generating a negligible difference in the value of the total thermal load.

4.5.1 Hypothesis 2 - shop located in Sao Paulo

Assuming locality variation for the commercial environment, a hypothesis was raised in which the shop is located in the state of Sao Paulo. The cooling loads varied, as shown in Figure 42.

This hypothesis arose because, due to the company's expansion project, it is being realised

analysing the thermal load in other locations so that, in the future, the total cost of adapting the establishment to the new standard can be estimated and, based on this, we can have approximate figures that will be invested in the air conditioning project for all the establishments that will start operating.

Figure 42 - Comparison of data by location.

	External		External
	A Condixjao Solar Inf. lat. Inf. sens. (m^2)(W)(W)(W) (W)		Solar conduction Inf. lat. Inf. sens. (m2) (W)(W) (W) (W)
Maximum cooling load per compartment Maximum cooling load per compartment			
Environment Shop Shop stock	473993720100 48819243000	Environment Shop Stock Shop	473682021700 48814858000
	R_e cife	Sao Paulo	

Source: Author.

5 CONCLUSION

When analysing the procedures used to study the thermal load for the feasibility of an air conditioning system in the room under study, the method proposed by the CYPETHERM LOADS software, which operates in an Open BIM workflow and meets the recommendations of the American Society of Heating Engineers (ASHRAE), together with the method proposed by ABNT NBR 16401, are highly effective, given their parameterisation using values already determined by tables.

The use of Open BIM workflow software provides very satisfactory results compared to other methods, from creating the IFC model of the structure to calculating the room heating hypotheses. It also allows the user to quickly calculate the thermal load and compare the results with different input parameters, making it an excellent engineering teaching tool.

In view of this, the values presented show that the air conditioning system adopted will need to be adapted in order to promote acceptable thermal comfort for the occupants of the room during business hours.

As a suggestion for future work, we could evaluate the sizing of an air-conditioning system capable of meeting the needs of the environment, as well as carrying out a cost survey for the installation, taking into account the cost-benefit ratio for the company, since this work details the total energy generated by this system.

REFERENCES

Brazilian Association of Technical Standards. **NBR 16401-1: Air Conditioning Installations - Central and Unitary Systems - Part 1: Installation Projects**. Rio de Janeiro, 2008.

Brazilian Association of Technical Standards. **NBR 16401-2: Air Conditioning Installations - Central and Unitary Systems - Part 2: Thermal Comfort Parameters**. Rio de Janeiro, 2008.

Brazilian Association of Technical Standards. **NBR 16401-3: Air Conditioning Installations - Central and Unitary Systems - Part 3: Indoor Air Quality**. Rio de Janeiro, 2008.

CENGEL, Yunus A.; BOLES, Michael A. **Thermodynamics.** 7 ed. Porto Alegre: AMGH, 2013.

CREDER, Helio. **Air Conditioning Installations**. 6ª Edition. Rio de Janeiro: LTC, 2004.

EASTMAN, Chuck; TEICHOLZ, Paul; SACKS, Rafael; LISTON, Kathleen. **BIM Handbook:** a construction information modelling guide for architects, engineers, managers, builders and developers. Porto Alegre: Bookman, 2014.

FOX, Robert W.; MCDONALD, Alan T.; PRICHARD, Philip J.; LEYLEGIAN, John C. **Introduction to fluid mechanics.** 8 ed. GEN: Sao Paulo, 2014.

FROTA, Anesia Barros; SCHIFFER, Sueli Ramos. **Manual of thermal comfort.** 5 ed. Sao Paulo: Studio Nobel, 2001.

INCROPERA, Frank P. **Fundamentals of Heat and Mass Transfer**. 6 ed. Rio de Janeiro: LTC, 2014.

INCROPERA, Frank P. **Fundamentals of Heat and Mass Transfer**. 8 ed. Rio de Janeiro: LTC, 2014.

INMETRO. **Balango Winter 2021 in the capitals Recife-PE and Joao Pessoa-PB** .2021. D i spo ni vel at: < Https://portal.inmet.gov.br/noticias/balan%C3%A7o-do-inverno-2021-nas-capitais-recif>.

JESUS, Marcos Fabio; SILVA, Gabriel Francisco da. Programme for estimating psychometric properties. **Revista Brasileira de Produtos Agroindustriais.** Campo Grande, v. 4, n. 1, p. 63-70, 2002.

MARCOS, Micheline Helen Cot. **Analysis of CO2 emissions in the pre-operational phase of social housing construction using a cad-bim tool.** Dissertation presented to the Postgraduate Programme in Civil Construction at the Federal University of Paraná. Curitiba, 2009.

MARQUES, Milton Cesar Silva; HADDAD, Jamil; MARTINS, Andre Ramon Silva. **Energy conservation:** energy efficiency of equipment and installations. 3 ed. FUPAI: Itajuba, 2006.

MORENO, Lucas Mesquita. **Design and sizing of ventilation networks and ducts for intensive care units.** Monograph
(Graduation)- Federal University of Ceara. Fortaleza, 2017.

VAROTTO, Luis Fernando. Retail in Brazil - a historical overview and trends. **Revista Brasileira de Marketing**, v. 1, p. 430, 2018.

SBVC. **The role of retail in the Brazilian economy.** 2021.

SILVA, Andreia de Oliveira. **Ergonomics in the workplace: a case study at SUPGA/SERPRO.** Monograph (Graduation) - Faculty of Applied Social Sciences, University Centre of Brasilia. Brasilia, 2005.

SPITLER, Jeffrey D.; COOK, Jack C.; LIU, Xiaobing. A Preliminary Investigation on the Cost Reduction Potential of Optimising Bore Fields for Commercial Ground Source Heat Pump Systems. 45th **Workshop on Geothermal Reservoir Engineering Stanford University**, Stanford, California, February 10-12, 2020.

WYLEN, Gordon Van. **Fundamentals of Classical Thermodynamics**. 4 ed. Sao Paulo: Edgard Bluncher, 2003.

CHAPTER 2

PLASMA NITRIDING OF TITANIUM NITRIDE THIN FILMS ON GLASS SAMPLES USING CATHODIC CAGES

SUMMARY

Nitriding is a common heat treatment process for increasing surface hardness and wear resistance, making it a process that offers many industrial advantages. Ion nitriding has come to offer faster and less time-consuming treatment of a part, making it possible to add a variety of materials due to the low treatment temperature and speed. As technology advanced, improvements were made to the nitriding process to avoid edge effects, overheating and layer uniformity. This led to the introduction of the cathodic cage plasma nitriding technique, in which the plasma draws electrons from the cage, which combine with other reactive elements and are deposited on the substrate at a fluctuating potential. The optical properties of TiN can be put to great use in solar panels as infrared filters. This paper presents a literature review on analysing the efficiency of the cathodic cage nitriding technique for depositing Titanium Nitride (TiN) and its influence on glass samples. Articles, dissertations and theses were read on thermochromic treatments with certain temperatures, modifications to the cage and specific pressure and duration conditions. Analyses were carried out in which the samples were characterised and compared according to their treatment conditions. The results obtained from these tests proved that a thin TiN film was deposited under all treatment conditions, proving the efficiency and great versatility of this deposition technique.

KEYWORDS: Plasma nitriding, Cathodic cage, Titanium nitride thin films, Plasma deposition. Glass.

1 INTRODUCTION

Nitriding is a common heat treatment process for increasing surface hardness, mechanical properties, corrosion resistance and wear. Many machine parts such as toothed gears, workpieces such as camshafts and tools such as screwdrivers are industrially nitrided to obtain improvements in their tribological and chemical properties (SCHAAF, 2002).

Ion nitriding is a process that offers many industrial advantages and is used to improve various physical properties of metal surfaces. It has several advantages over gaseous or salt bath nitriding processes, but there are some disadvantages, but several alternatives have already been created to eliminate or alleviate them.

Steel is a material that is widely used in the most diverse sectors of industry, but many have poor tribological properties. Tribology is the science that studies the friction, wear and lubrication that takes place during contact between solid surfaces in relative motion, hence the indication of treatments that are very effective.

Surface nitriding treatments on steel are a growing reality, but in general, questions arise about using plasma nitriding treatment on other materials. For example, if glass is subjected to plasma nitriding treatment, what can be expected as a result? Will a film be deposited? And what use will industry make of this new material? In a way, that's what we want to do with this bibliographical research: discuss the deposition of titanium nitride films on glass samples.

Titanium nitride films can be obtained using this technique and have many uses in industry, such as increasing surface hardness and reducing friction wear (PENG et al., 2003), decorative coatings to replace gold, and in glass, it can be used as a coating for solar panels, solar control windows (SMITH; BEN-DAVID; SWIFT, 2001), biomaterials (CHENG; ZHENG, 2006) and microelectronic semiconductors (WITTMER; NOSER; MELCHIOR, 1983).

Bearing in mind that the usability of TiN is very high, the present work aims to carry out a bibliographical review of surface treatments on glass samples using a titanium cathode cage, under varying conditions of temperature and treatment time for the same ratio of the N2-H2 gas mixture, in order to answer the question: what is the influence of titanium nitride deposition on glass? and also to evaluate the properties of the samples after nitriding.

To do this, we need to carry out a systematic study of the various nitriding parameters, such as temperature and duration of the process, as well as measuring the depth of the nitrided layer using scanning electron microscopy, identifying the type of film deposited using XRD and SEM, and, if possible, checking the electrical conductivity using a conductivity meter.

The current work was carried out following the criteria of basic research, bibliographical type, with a qualitative approach and is divided into 4 sections with 15 subsections in order to facilitate understanding and concisely present the most relevant elements of this research. The exploratory objective of the work was successfully achieved through bibliographical research.

Articles, master's theses, doctoral dissertations and books were selected according to the search terms already described in the search fields of the academic work repositories. The criteria considered when analysing the scientific articles were: concepts covered, types of nitriding, surfaces used, influence on temperature, data collection equipment.

Within the broad subject of nitriding, the delimitation took the form of the application of the technique to glass samples, more specifically, the deposition of thin titanium nitride films by plasma nitriding on glass samples using a cathodic cage. In order to achieve the application of the technique, the chapters have evolved from what is meant by plasma, followed by nitriding, plasma nitriding with cathodic cage to the treatment of glass samples.

2 LITERATURE REVIEW

2.1 PLASMA

In relation to the concept of plasma, Howatson (1976) uses the term *plasma* to describe a wide variety of macroscopically undefined substances containing many interactions between free electrons, atoms, radicals and molecules, both neutral and ionised, which exhibit collective behaviour motivated mainly by Coulomb forces. This concept is further expanded by emphasising that not all of the plasma volume contains charged particles, but it can be classified as plasma.
Alves Junior (2001) adds that normally a plasma is electrically neutral and any charge imbalance will result in electric fields that tend to move the charges in order to re-establish equilibrium. As a result, the density of electrons plus the density of negative tones must equal the density of positive tones.
Because of this phenomenon, an important plasma parameter is called the degree of ionisation, which represents the fraction of the original neutral species that have been ionised. Alves Junior (2001) also adds that there is a plasma with a much lower degree of ionisation than the unit called cold plasma, which is used in the ionic nitriding process.
According to the same author, plasma can be produced when there is a potential difference between two electrodes contained in a system with sufficiently low pressure and hermetically sealed. Electrons and ions are accelerated by the electric field colliding with other particles and produce more tones and electrons through the following chemical equation:

$e- + G0 \rightarrow G^{+} + 2e-$

Where G^{0} is the atom or molecule of the gas in the ground state and G^{+} represents an ion of this gas.

2.2 IONIC NITRIDING

Pinedo (2004) explains in general terms that nitriding is a term used to refer to techniques in which nitrogen is introduced to the surface of a material in order to improve its surface properties. One of the most prominent nitriding processes is that which uses plasma as an energy source and is called plasma nitriding or ionic nitriding. This process, patented by J.J. Egan in 1931 in the USA and by Berghaus in 1932 in Germany, only began to be used commercially in the 60s, with great progress in the 70s.
The technique or process of ionic nitriding or plasma nitriding is based on a hermetically sealed system - reactor, with a voltage difference between the

cathode (the sample holder) and anode (reactor housing), with a vacuum pump to achieve a pressure of around 10^{-2} torr and a valve to control the flow of gases as shown in figure 1.

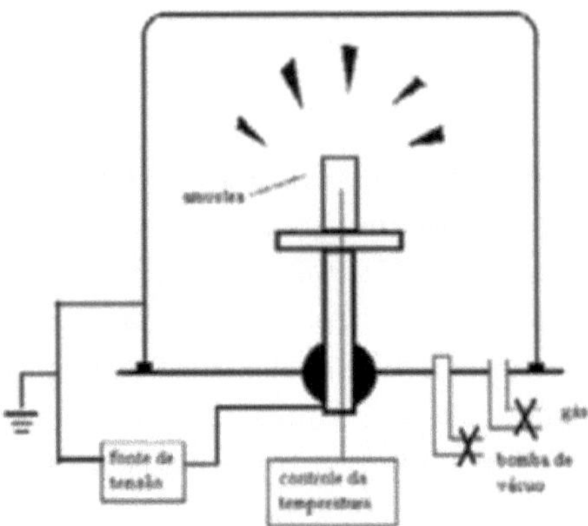

Figure 1 - Basic diagram of ionic nitriding equipment (ALVES JR, 2001)

In the case of ionic nitriding, the sample to be nitrided is placed in the sample holder and initially a vacuum of around 10^{-2} torr is produced in the reactor. A potential difference of between 400 and 1200 V is applied between the electrodes and then the nitriding gas (typically a mixture of N_2/H_2) is introduced into the reactor until the working pressure is reached. According to Pinedo (2004), when the potential difference is applied between the cathode (handles) and the anode (furnace casing), in the presence of the gas mixture, under conditions of spectacular temperature and pressure, a *glow* discharge is generated, which determines the occurrence of the plasma (figure 2). In this process, the gaseous molecules are dissociated, the positively charged tones are accelerated towards the surface of the anode and the electrons are directed towards the cathode.

Figure 2 - Formation of bright discharge (plasma sheath) in nitriding (NITRION BRASIL, 2019)

During nitriding, a luminous region called cathode luminescence arises due to the excitation of surface atoms bombarded by plasma species. Between the iris of this luminescence and the cathode there is a dark space called the cathode sheath, which is a region of low charge concentration due to the potential gradient. After this dark space there is a region of high luminosity called negative luminescence. The negative luminescence, together with the cathode region (cathode and sheath), is responsible for the release of electrons and some phenomena such as charge transfer, ionisations and excitations and the production of secondary electrons (ALVES JR, 2001). In a nitriding medium (N_2/H_2 mixture), the most frequent reactions can be classified as follows:

Ionisation - When the more energetic electrons collide with the gas molecules, they remove an electron from the atom, producing one ion and two electrons:

$$e^- + N_2 \longrightarrow 2e^- + N_2^+$$

$$e^- + H_2 \longrightarrow 2e^- + H_2^+$$

$$e^- + N_xH_y \longrightarrow 2e^- + N_xH_y^+$$

Excitation - If the collision energy of the electron is lower than that required for ionisation, the following excitations can occur

$$e^- + N_2 \longrightarrow e^- + N_2^*$$

$$e^- + H_2 \longrightarrow e^- + H_2^*$$

$$e^- + N_xH_y \longrightarrow e^- + N_xH_y^*$$

where * represents the excited state.

Relaxation or emission - After the excitation of the species, the electrons decay to less energetic levels, resulting in the emission of photons:

$N_2^* \longrightarrow N_2 + hv$

$H_2^* \longrightarrow H_2 + hv$

$N_xHy^* \longrightarrow N_xH_y + hv$

By analysing the light emitted by the discharge, it is possible to diagnose the plasma.

Dissociation - Another important reaction that takes place in plasma when molecules collide with energetic electrons is dissociation. Depending on the energy of the electron, neutral, excited or ionised atoms can be formed, in the particular case of nitrogen:

$e- + N_2 \longrightarrow e^- + N + N$

$e- + N_2 \longrightarrow e^- + N^* + N$

$e- + N_2 \longrightarrow e^- + N^* + N^+$

Recombination - When ionised species collide with a surface, electrons from that surface are released, neutralising the species through the following recombination processes:

$N_2^+ + e^- \longrightarrow N_2$

$H_2^+ + e^- \longrightarrow H_2$

$N_xH_y^+ + e^- \longrightarrow N_xH_y$

The plasma nitriding technique can be applied to various types of materials. In the use of tools, wear is one of the most important factors leading to the end of tool life. Therefore, the life of a tool and its integrity depend mainly on the mechanical properties of its surface, and nitriding is one of the processes used to optimise tool life. This process has advantages such as, according to Alves Junior (2001), reduced treatment time and temperature, better control of the nitrided layer, uniform layer thickness, nitriding in parts of the tool, the possibility of denitriding, and lower energy costs compared to other nitriding processes.

Alves Junior (2001) also mentions, for example, the hollow cathode effect, the effect of the A/V (area/volume) ratio, which causes the sample to overheat, the opening of cathodic arcs in handles with more complex geometries, the difficulty of penetrating small holes and the slow heating rate if you want a stable plasma without the risk of cathodic arcs. All these limitations can either extinguish the plasma or jeopardise the treatment.

In recent years, major advances have been made in the design of equipment in the areas of power generation, automation, handle heating methods and cathodic cage nitriding techniques which, in addition to eliminating previously existing problems, have also improved the operational and economic aspects of the process.

2.3 PLASMA NITRIDING WITH CATHODIC CAGE

To solve the limitations of ionic nitriding LI, C. X. et al.,2002, proposed a nitriding technique called *Active Screen Plasma Nitriding* (ASPN) which consists of adding a metal screen with no defined geometry to be the cathode while the handles to be treated remain in the sample holder under an electrical insulator to remain at zero potential. In this way, the plasma acts on the screen and not on the surface of the samples, thus avoiding edge effects and surface wear, as shown in figure 3:

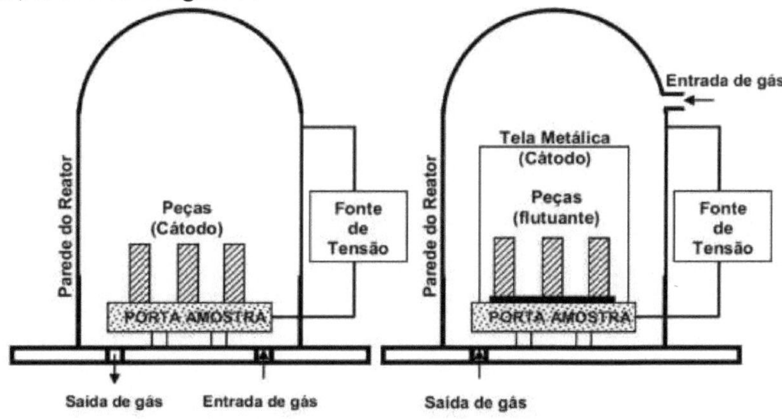

Figure 3 - Schematic diagram showing a system (a) ionic nitriding (b) ASPN (LI, C. X. et al., 2002)

Thanks to the efforts of LI, C. X. et al. (2002), it was possible to investigate an even more efficient device than the active screen nitriding process, called the Cathodic Cage, in order to obtain defect-free surface treatments that could be used in a wide range of industrial applications.

This new technique uses a metal screen with a well-defined geometry, i.e. it has holes with a well-defined and standardised diameter and distance between the holes, so that multiple hollow cathodes can be used simultaneously.

According to Sousa (2007) in comparison with LI, C. X. et al. (2002), ASPN treatment over 20 hours on SAE 316 stainless steel resulted in nitrided layers 7 ^m thick while plasma nitriding with a cathodic cage, treating the same steel at the same temperature (450 °C), resulted in layers 18 ^m thick in just 5 hours of nitriding.

The cathode cage consists of a cylindrical plate with holes in it and a circular plate with holes in it to be the lid.

The principle is based on the Faraday Cage experiment, in which the handle is placed on top of an electrical insulator and surrounded by the cathodic cage so that its electric field inside is zero and the electrons are distributed in the outer region of the cage.

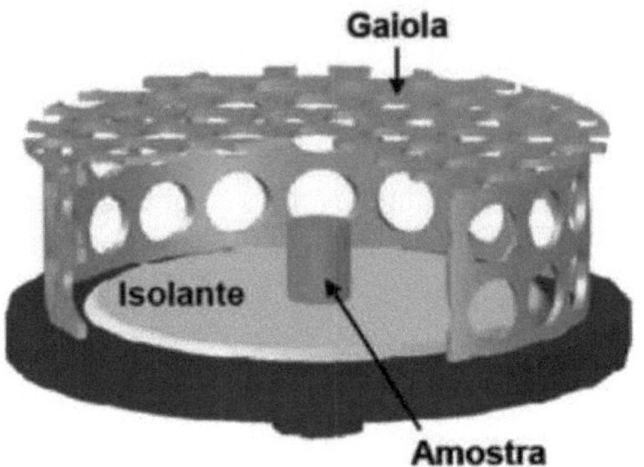

Figure 4 - Cutaway view of the cathodic cage (ARAUJO, 2006)

Figure 5 shows the visual aspect of the plasma formed on the cage. According to Araujo (2006) and Daudt (2012), it is possible to observe the luminous intensification of the plasma in each hole, this effect being typical of a hollow cathode with a luminous direction out of the holes in the cage similar to multi-cylindrical cathodes due to the difference in potential between these regions. They add that this effect depends on the hole diameter and the working pressure, occurring at a specific pressure for each hole diameter.

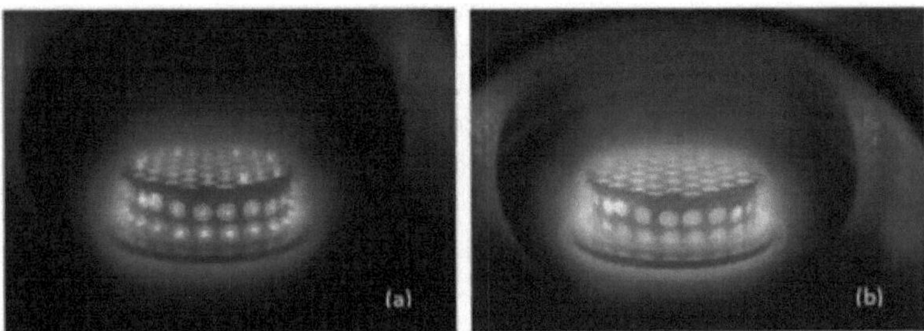

Figure 5 - Visual aspect of plasma formation on the cage surface as a function of working pressure: (a) P = 2.5 mbar and (b) P = 5.0 mbar (ARAUJO, 2006)

Silva et al. (2017) explains that the plasma acts on the cathodic cage and pulls material, mainly from the walls of the holes, which reacts with plasma gases, in this case nitrogen, and deposits on the substrate positioned inside the cage in suspended potential.

Hollow cathode discharges occur in the holes in the walls of the cage, producing higher densities of tones than conventional discharges, because the density of the plasma inside the holes is much higher, which is why a high quantity of tones is ejected from the walls of the holes. According to Koch et al, maximum scattering occurs when the pressure x cavity size product varies approximately between 0.375 and 3.75 cm.Torr.

Sousa (2007) concludes through comparisons with ionic nitriding that cathodic cage nitriding achieves the same properties, but eliminates the common problems associated with conventional treatment, especially the edge effect, providing uniform three-dimensional layers, as exemplified in Figure 6.

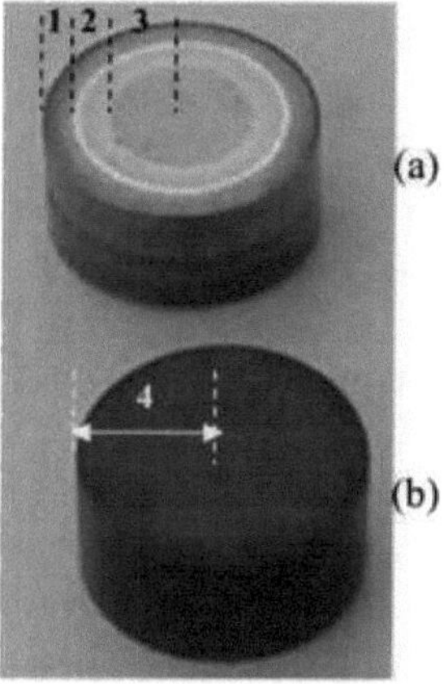

Figure 6 - Photo of AISI 420 nitrided by (a) ionic nitriding and (b) cathodic cage nitriding, respectively. (SOUSA, 2007)

It is known that for metallic samples, the elimination of the edge effect, which is a characteristic of processes using the Cathodic Cage (SOUSA, 2007), only occurs when the sample is nitrided at floating potential, i.e. positioned on the insulating disc, because the deposition covers the entire inner surface of the cage base.

2.4 TIME AND TEMPERATURE IN THERMOCHEMICAL TREATMENT

Martensitic stainless steels, for example, are sensitive to the nitriding temperature because they have low resistance to tempering. The combination of nitriding temperature and time is essential to guarantee surface hardening without loss of core hardness. It is also known that the nitriding temperature can significantly affect the corrosion resistance of these steels due to the precipitation of chromium nitrides in the nitrided layer.

According to Dos Reis et al. (2008), who studied the influence of treatment time on the plasma nitriding of ISO 5832-1 austenitic stainless steel, they obtained results that showed non-diffusive growth of the composite layer.

They realised that the increase in time, from 1 to 3 hours, promotes greater incorporation of nitrogen into the surface of the material, leading to the formation of phases richer in this element, as well as an increase in hardness and mass gain, and growth of the nitrided layer. Processing for 3 and 5 hours showed similar results.

Wanke et al. (2000) studied the influence of nitriding temperature on the formation of nitride layers in conventional and sintered M2 quick-sintering. They studied the influence of temperature on the properties (thickness and microhardness) and characteristics of the nitride layers. To do this, they nitrided samples at temperatures of 400 and 500 °C for 2 hours in an atmosphere composed of 80% N_2+ 20% H_2.

The results show that the thickness of the nitride layer increases with temperature and that the microhardness of the layer is around 50% higher than that of the substrate for both conventional and sintered auger. Further studies have been carried out in relation to the treatment temperature and all come to an approximate conclusion.

This is seen in the work of Bortoli et al. (2012), who studied the influence of plasma nitriding temperature on the thickness and wear resistance of expanded austenite layers formed on AISI 316L stainless steel. The thermochemical treatments were carried out using gas mixtures of 75% N2 and 25% H_2 for 6 hours at varying temperatures of 410, 420 and 430 °C.

X-ray diffraction showed that all conditions produced nitrided YN layers with thicknesses of 3.5 - 6 iim. According to the study, the wear resistance of the YN layers tended to decrease as the treatment temperature increased, indicating the possible precipitation of phases harmful to the wear resistance of the auger.

AISI 316L austemitic stainless steel nitrided under a 75%N2 + 25%H2 gas atmosphere for 6 hours shows maximum resistance to abrasive wear when nitriding treatment temperatures of 410 °C are used. The nitrided layer generated in this condition has a thickness of approximately 3.6 inn and consists exclusively of expanded austenite (YN).

Although this layer is thinner than those obtained at temperatures of 420° C and 430° C, it has a higher wear resistance, since the material's wear coefficient drops from $1.50x10^{-12}$ m^2 /N to $0.99x10^{-12}$ m /N.^{2}

For the authors Prado et al. (2012), who studied the effect of plasma nitriding temperature on the microstructure of AISI 420 martensitic stainless steel, the effects of nitriding temperature can significantly affect the corrosion resistance of these steels due to the precipitation of chromium nitrides in the nitrided layer and that at high temperature the nitrided layer is hardened by the precipitation of iron and chromium nitrides, on the other hand, a decrease in the nitriding temperature inhibits the precipitation of chromium nitrides and promotes the formation of expanded martensite with the precipitation of iron nitrides.

They used AISI 420 Martensitic Stainless Steel and concluded that the use of

different plasma nitriding temperatures promotes the formation of nitrided layers with different microstructural characteristics, hardening potential and corrosion behaviour, i.e. the use of the low temperature cycle, 380 °C, promotes the formation of a nitrided layer composed of "expanded tempered martensite", "a'N"-cubic.
At the same time, iron nitrides of the e-Fe3N and g'-Fe4N types precipitate. There is no precipitation of chromium nitrides, while at high temperature (550 °C) this does not occur. The hardening potential is different for the different nitriding conditions. At low temperature, the hardening potential is lower than at high temperature, but it is still capable of raising the original hardness of the substrate from 600 HV to values close to 1,000 HV. At low temperatures, hardening is caused by the expansion of the tempered martensite lattice and the precipitation of iron nitrides, and it is not possible to isolate the contribution of each phenomenon. At high temperatures, the hardening potential is substantially greater, raising the hardness to around 1,400 HV as a result of the intense precipitation of nitrides in the nitrided layer, mainly chromium nitrides.
The corrosion resistance of the nitrided samples, assessed by measuring polarisation resistance, is reduced compared to the condition without nitriding treatment; however, the deleterious effect introduced by this treatment is less after nitriding in the low temperature condition, probably due to the absence of chromium nitrides in the nitrided layer after this treatment.

2.5 THIN FILMS

A thin film is a layer with a very thin thickness compared to the other dimensions of a given material deposited on a substrate, in the order of micrometres (WASA 2004, apud DAUT 2012, p. 28). They can also be defined as materials made by depositing particles (atoms, molecules, tones) on a substrate.
The use of technology to produce films is present in various sectors, from engineering to medical sciences and even decorative practice (COSTA, 2014).
The formation of thin films requires three processes (WASA, et al., 2004): 1. adequate production of atomic, molecular or ionic species; 2. transport of these species to the substrate through a medium; 3. condensation on the substrate, either directly or through a chemical and/or electrochemical reaction, to form a solid deposit.
Films are normally formed from the solidification of a vapour, which usually comes from thermal evaporation and/or evaporation of the source material on the substrate. The condensation process begins with the formation of small clusters of material, called nuclei, scattered randomly over the surface of the substrate.

Electrostatic attraction forces are responsible for attaching the atoms to the surface. The attachment mechanism is called chemical adsorption when electron transfer occurs between the substrate material and the deposited particle, and physical adsorption if this does not occur. The binding energy associated with chemical adsorption ranges from 8 eV to 10 eV and that associated with physical adsorption is approximately 0.25 eV. Adsorbed atoms migrate over the surface of the substrate interacting with other atoms to form nuclei. The process is called random nucleation, in which the nuclei grow and form larger clusters until they form islands.

This is followed by the coalescence process in which the small islands begin the process of engulfing others until they form a completely continuous film. Figure 2.9 shows the sequence of stages during formation.

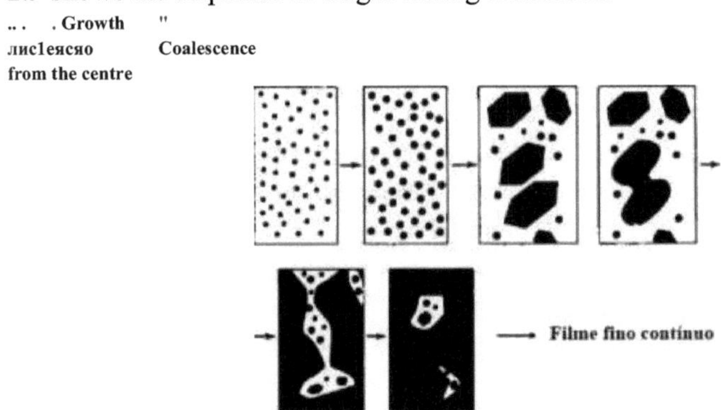

Figure 7 - Schematic representation of the sequence of stages during the formation of a film (TENTARDINI, 2004)

Currently, surface treatments using thin film deposition are of great importance for applications in the metal-mechanical industry as protection for substrates subjected to mechanical stress such as mechanical wear and corrosion (TENTARDINI, 2004), as well as decorative coatings, in addition to being used in medicine and dentistry as highly biocompatible coatings for implants (CHUNG, 2004).

Another very obvious use is in microelectronics and optical coating, which are applied for anti-reflective effect, in interference films in solar panels, glass sheets that reflect infrared and laser optics.

2.6 TITANIUM NITRIDE (TiN)

According to Chung et al. (2004), in recent decades TiN thin films have been widely used in medicine and dentistry as highly biocompatible coatings for implants. Tentardini (2004) adds that in addition to the biomedical industry, the metal-mechanics industry has been studying, developing and applying TiN films

as coatings to protect substrates subjected to mechanical stress such as mechanical wear and corrosion.
In fact, TiN films were initially developed for application as a coating on cutting tools in order to improve surface properties such as hardness, corrosion resistance, wear resistance and consequently extend the life of these tools. It is still widely used today for this purpose (PENG, et al., 2003).
TiN is a refractory compound that combines ceramic properties such as high melting point, high hardness, thermal and chemical stability, resistance to wear and corrosion, with some metallic properties such as low friction coefficient and high thermal and electrical conductivity (FOUILLAND, et al., 1998).
TiN films were initially developed for use as coatings on cutting tools in order to improve surface properties such as hardness, corrosion resistance, wear resistance and, consequently, extend the life of these tools. It is still widely used today for this purpose (PENG, et al., 2003).
Smith et al (2001) obtained TiN thin films with a high Ti/N ratio (1:3) without destroying the characteristic metallic properties of stoichiometric TiN. According to them, the change in electron mobility density shows that by depositing thin films, the irradiance in the increase in reflectance was shifted to the region neighbouring the near infrared; this makes it possible to produce films capable of transmitting white daylight at reasonably high levels, while maintaining solar control in the near infrared and a low emittance.
This can be applied to solar control windows, as it uniformly transmits the components of visible light and blocks infrared and ultraviolet radiation from the sun by reflection. Yuste et al (2011) also obtained titanium nitride-based coatings with high transmittance and low emissivity, proving their high potential for use in solar control windows and solar thermal collectors.
Since the 1980s, titanium nitride has also been studied as a coating for semiconductors. Its low electrical resistivity (55^Qcm) compared to pure Ti films (65^Qcm), as well as its good metallurgical and chemical stability, make it very attractive for use in semiconductor devices (WITTMER, et al., 1981 and WITTMER, et al., 1983).
Composite films can be obtained by introducing reactive gases into a chamber. TiN films can be obtained from a Ti target and a hydrogen and nitrogen plasma.

2.7 TiN DEPOSITION ON GLASS SAMPLES

Titanium nitride (TiN) thin films are widely used in industry. Silva et al. (2017) explain why this usability is due to the high hardness, high melting point, good resistance to wear and corrosion of TiN films.
The authors add that because of these characteristics, it is often used as a coating on cutting tools such as milling cutters and drills, which acquire high hardness

and low friction wear, making them have a longer service life and good cutting performance. Yuste et al. (2011) and Zheng et al. (2007) state that TiN films are a more viable option for filtering infrared light due to their high reflection of this light band.

2.7.1 The glass used in deposition

For the deposition on glass Silva et al. (2017), Daudt (2012), Fernandes et al. (2018) used borosilicate sheets with dimensions of 25 mm by 10 mm with different thicknesses of each, respectively, 1 mm, 2 mm, 1.2 mm while Sousa et al. (2015) used glass samples with dimensions of 20 mm x 20 mm x 2 mm and composed of 72% Silica dioxide (SiO_2), 14% Sodium Oxide (Na_2O); 9% Calcium Oxide (CaO); 4% Magnesium Oxide (MgO); 0.7% Aluminium Oxide (A^O3) and 0.3% Potassium Oxide (K_2O).

2.7.2 Preparing the glass samples

Preparing the samples before starting the treatment is necessary to avoid contaminating the sample in the final results, thus resulting in the contaminated sample being disregarded. Silva et al. (2017) cleaned the substrates with detergent, water, an ultrasound bath for 1 minute and then dried them.

After drying, the substrates were placed inside the cage. Daudt (2012) sanded the edges of the samples with SiC water sandpaper with grain sizes of 100, 220 and 400. After sanding, the samples were cleaned in an ultrasonic p.a. acetone bath for 10 minutes and dried with a 60 °C air current.

Sousa et al. (2015) carried out two cleaning processes, first with a KOH solution followed by an HNO3 solution and then in an acetone bath using ultrasound. Once the samples had been prepared, they were positioned under the electrical insulator that was placed in the sample holder, the side of the cathodic cage was placed so that the glass slides were closer to the centre of the cage and then the cathodic cage lid was installed.

2.7.3 Characteristics of cathodic cages

As explained in previous sections, in cathodic cage nitriding the nitriding atmosphere pulls tones or molecules out of the cathode, in this case the cathodic cage, and for the deposition of TiN on the glass a cage made of titanium is needed so that titanium atoms are pulled out and combine with the reactive nitrogen in the atmosphere to form the TiN compound that will be deposited on the substrate, diffusing and forming the film.

Silva et al. (2017) used commercially pure grade II titanium cages with an external diameter of 100 mm with a 45 mm high side configuration with 3 rows of holes and several lids with 37 9 mm diameter holes equidistant by 12.9 mm.

Daudt (2012) used 4 cages with 4 lids, 70 mm in diameter, 1 mm thick, made of commercially pure grade II titanium with 2 different hole diameters of 8 mm and 12 mm. For the 8 mm diameter, the distance between the centre of the holes was 9.2 mm and for the 12 mm hole diameter, the distance was 13.2 mm.

Sousa et al. (2015) used two 2 mm thick commercially pure grade II titanium cages for double deposition. The outer cage measured 75 mm x 55 mm (diameter and height) and the inner cage 45 mm x 35 mm, both with 8 mm holes equidistant by 9 mm.

2.7.4 Cleaning the cathode cages

The cage usually has to be cleaned before being placed in the plasma reactor to avoid contaminants in the final results, thus avoiding unexpected results.

Daudt (2012) and Sousa et al. (2015) mention the process of chemical cleaning. The first immersed the cages in an acidic solution containing plain water (33% by volume of deionised water, 33.3% by volume of HCl and 33.3% by volume of HNO_3) for 12 hours, followed by sanding with 220 and 600 grit sandpaper. The second author first sanded and polished the cages, then immersed the cages in a solution containing 50 ml of HNO3, 25 ml of HF and 425 ml of distilled water in an ultrasonic cleaner for 10 minutes, after which the cages were cleaned with acetone and dried with a hairdryer.

2.7.5 Preparing the reactor for treatment

After cleaning the glass samples and the cathode cages, the reactor is cleaned for the same reasons as before, and according to Daudt (2012) it can be mechanically cleaned using sandpaper in order to remove residues from previous processes.

Sousa et al. (2015) and Silva et al. (2017) do not mention mechanical cleaning like the previous author. After mechanical cleaning, the glass samples are positioned so that an electrical insulator is under the cathode and the glass samples are under the insulator. Afterwards, there is a second cleaning stage. Daudt (2012) used H2 plasma at a temperature of 200 °C, with an H2 flow of 8 sccm for 30 minutes.

Sousa et al. (2015) differentiated by cleaning the reactor with argon injection and subsequently evacuating it. This operation was repeated twice. Prior to deposition, the sample surfaces were *pre-sputtered* to remove oxides.

2.7.6 Plasma nitriding treatment with cathodic cage

Plasma treatment with a cathodic cage begins when all of the above is completed and the working temperature and pressure are reached. Daudt (2012) used a temperature of 450 °C for 120 minutes with different atmospheres of Air, N2 and H_2 as nitriding parameters, the flow of H_2 was the variable studied, so 0 sccm, 1 sccm and 2 sccm were used, while the other two gases had flow values of 4 sccm for Air and 3 sccm for N2 .

The same author justifies the use of gases: "Argon was used because of its known *sputtering* power. Nitrogen was used to promote the titanium nitride formation reaction and hydrogen was used as a 'catalyst' due to its ease of ionisation." Tables 1, 2 and 3 below were produced by the author to demonstrate the variables in each nitriding process studied:

Table 1 - Process variables with H2 flow = 0 sccm

Cage	Pressure (mbar)	Current (A)	Voltage (V)	Power (W)
LIT F48	2,36	0,31	815	252,6
LIT F88	2,38	0,32	822	263,1
L T F248	2,40	0,37	812	300,4
brTвГв	2,37	0,35	795	278,2
LIT F412	2,35	0,30	877	263,1
LIT F812	2,38	0,33	868	286,4
L2T4F12	2,40	0,35	846	296,1
L2T8F12	2,35	0,35	890	311,5

Source: (DAUDT, 2012)

Table 2 - Process variables with H2 flow = 1 sccm

Cage	Pressure (mbar)	Current (A)	Voltage (V)	Power (W)
L^Fe	2,46	0,30	783	234,9
LIT F88	2,45	0,30	805	241,5
L T F248	2,44	0,34	803	273,0
L T F 288	2,49	0,35	824	288,4
LIT F412	2,43	0,30	860	258
LIT F812	2,47	0,31	820	254,2
L T F2412	2,46	0,33	811	267,6
L T F 2812	2,49	0,36	878	316,1

Source: (DAUDT, 2012)

Table 3 - Process variables with H2 flow = 2 sccm

Cage	Pressure (mbar)	Current (A)	Voltage (V)	Power (W)
LIT F48	2,53	0,31	788	244,3
LIT F88	2,53	0,32	800	256
L T F248	2,58	0,36	785	282,6
L T F288	2,54	0,36	800	288
L1T4F12	2,53	0,32	808	258,6
LIT8 F 12	2,47	0,31	820	254,2
L2T4F12	2,54	0,35	817	286
L T28 FI2	2,54	0,36	828	298,1

Source: (DAUDT, 2012)

Sousa et al. (2015) used temperatures of 350 °C and 450 °C, with a duration of 1.5, 2 and 3 hours, a working pressure of 150 Pa and a flow of 12 sccm of H2 and 4 sccm of N2/H2 (80% N2).

While Silva et al. (2017) described their treatment using a vacuum pump, which was activated and the pressure reduced to 1 Torr, hydrogen was then introduced into the reactor chamber at a flow rate of 250 sccm. The reactor was switched on and the plasma formed on the cathode with the cage.

The system was heated by the plasma itself using hydrogen for the first 30 minutes at a pressure of 1.0 Torr and a flow of 250 sccm. After this time, nitrogen was introduced. At this point, the flow rates of the two gases into the reactor chamber were adjusted as shown in Table 4. When the temperature reached 300 °C, the plasma was kept operating in the cage for a further 1 hour. The following table shows the parameters of the 3 results chosen by the author for the analyses.

Table 4 - Parameters used at a temperature of 300 °C

Essay	H flow$_2$ (sccm(%))	Flow of N2(sccm(%))	Pressure (Torr)	Pressure x hole diameter (cm. Torr)
1	162 (64%)	88 (36 %)	1,0	0,90
2	28 (11 %)	222 (89 %)	1,0	0,90
7	95 (38 %)	155 (62%)	0,5	0,45

Source: (SILVA et al., 2017)

2.7.7 Analysing instruments

Sousa et al. (2015) reported that the equipment used to analyse composition and texture was Raman spectroscopy (laser 785 - Perkin Elmer) obtained using a Bruker Senterra single-grating spectrometer equipped with a charge-coupled device, as well as a detection system (CCD).

A 785 nm laser was used for excitation. The lenses of an *Olympus* microscope with a focal length of f = 20.5 mm and a numerical gap of Na = 0.35 were used to focus the laser beam on the surface of the sample. Gaps in the spectrometer were created to obtain a resolution of 3 cm^{-1} . The laser power was approximately 25 mW at the sample surface and the recording time was 20 seconds with twenty accumulations.

The author also used an X-ray diffractometer (XRD) and his analyses were carried out using Cu Ka lines with a wavelength of 0.154 nm operating with a Shimadzu XRD-6000 40 KV x-ray diffractometer.

To observe the morphology of a nitrided layer, an *Olympus* BX60M microscope coupled with a digital camera and *Image Pro - Plus* was used. Layer thickness was observed using scanning electron microscopy. A Shimadzu SSX-550 model, coupled with EDS, was used to identify the chemical composition of the deposited film, as well as to visualise and measure the thickness of the deposited nitrided layer in cross-sections.

The equipment that Daudt (2012) used to collect data was an optical emission spectroscopy system to study the influence of the active species present in the plasma on the microstructure and optical properties of titanium nitride films. It consisted of an *Ocean Optics* USB 4000 spectrograph with an optical resolution of 0.3 to 10 nm, a local length of 42 mm at the entrance and 62 mm at the exit and an optical response between 200 - 1100 nm. It has a Toshiba TCD1304AP

linear CDD series detector. An *Ocean Optics* HR4000CG spectrograph was also used to resolve some of the peaks.
The shallow angle X-ray diffraction measurements by Daudt (2012) were carried out in the laboratory of the Nucleus of Studies in Petroleum and Natural Gas (NEPGN) at UFRN and the microanalysis laboratory of the Institute of Physics at UFRGS, using a SHIMADZU XRD-6000 diffractometer, with thin film accessory, with Cu Ka radiation. A 20° scan from 30° to 45° was used, since the formation of peaks is concentrated in this range.
Daudt's (2012) optical microscopy was carried out using an *Olympus* model BX60M with an *Express - Series* camera attached and the image was captured using *Image Pro - Plus* software. The analysis was carried out in order to check the homogeneity and surface characteristics of the film.
Daudt (2012) differs from Sousa et al. (2015) and performs a UV/VIS spectrophotometry analysis to characterise the optics of the samples through absorbance, transmittance and reflectance and emphasises that the measurements were taken using air as a blank. The transmittance measurements were made using a *Genesys* 10 UV spectrophotometer, with a xenon lamp, capable of scanning the wavelength range between 190 and 1100 nm, with an accuracy of ±1 nm, with a spectral bandwidth of 1.8 nm, with an optical path of 10 nm, with a transmittance reading in the range of 0.3 to 125 %T. All measurements were taken at the same height as the sample. The transmittance and total reflectance measurements for some of the author's samples were made using another Varian Inc Cary-5000 spectrophotometer in the Laser and Optics Laboratory at the UFRGS Institute of Physics. This device has tungsten and deuterium lamps with an accuracy of 0.0006 nm. It scanned from 350 nm to 1100 nm.
The transmittance measurements for monochromatic light by Daudt (2012) were made using a monochromatic laser, a red HeNe laser with a wavelength of 632.8 nm. An Acton Spectrapro 2500i emission spectrometer with a focal length of 500 nm and a maximum focal resolution of 0.05 nm was used to acquire the values.
Daudt's (2012) ellipsometry readings were taken using a Sopra GES-5E ellipsometer, which consists of a xenon lamp, a rotating polariser, a fixed polariser analyser and a spectrometer, thus enabling the optical constants of thin films to be obtained in the spectral range from 0.25 to 1.88 ^m.The author explains that it is a non-destructive analysis technique that makes it possible to measure the refractive index (*n*) and extinction index (k) and the thickness of thin films by changing the polarisation state of collimated beams of polarised monochromatic light caused by reflection from surfaces.

Daudt (2012) used atomic force microscopy with a SHIMADZU model SPM 9600 microscope. It basically consists of a test tip that scans the surface of the sample. The interaction force between the atoms on the tip and those on the surface is measured and, using computer resources, the results are transformed into images of the sample surface in three dimensions. The mode used to scan the sample was contact mode, obtaining images with areas of 1 micron, analysing roughness values.

Silva et al. (2017) demonstrated their data collection through X-ray diffraction (XRD), which was carried out at the Propem/Ifes/Vitoria Laboratory using a Bruker model D2 Phaser diffractometer, Cu-ka radiation (=0.1542 nm), at 30kV and 10 mA. The scan (2 from 20 to 80°) was made with a step of 0.01° per second. X-ray reflectometry (XRR) was used to determine the thickness, roughness and density of the films and is an analytical technique that investigates the structures of thin layers of surfaces and interfaces using the effect of the total external reflection of x-rays. For this purpose, the Bruker D8 Discover X-ray diffractometer belonging to the Physics Department of the Federal University of Vigosa was used, as well as Cu Ka radiation and a 0.002° pass, 1 s per step, 40 KV / 30 mA.

The authors differ from Sousa et al. (2015) and Daudt (2012) in their search for the electrical properties (resistivity, mobility and number of charge carriers) of the samples, which were measured at room temperature using an Ecopia model HMS - 3000 Hall Effect apparatus with a magnetic field of 0.556 T belonging to the Propemm/Ifes/Vitoria Surface and Thin Film Laboratory.

The four-point method was used, with square samples and a distance of 1 cm between the points. The transmittance curves of the films were determined using a Variant Cary 1E spectrophotometer belonging to the UFES Chemistry laboratory. The scans were made at wavelengths in the 300 nm to 1000 nm range. Borosilicate glass was used as a blank.

Scanning electron microscopy and an energy dispersive system (SEM and EDS) were used to characterise the films and their compositions, using a ZEISS model EVO MA10, belonging to Propemm/IFES. In addition, the images and analyses were used to compare, verify and quantify the presence of titanium and nitrogen in the deposited films.

3 RESULTS FOUND

This section will be used to present and discuss the results found in the publications by Daudt (2012), Sousa et al. (2015) and Silva et al. (2017).

With the aim of double deposition, Sousa et al. (2015) looked for indications of deposition through Raman spectroscopy, X-ray diffraction and EDS, in this article they did not try to investigate other characteristics of the TiN films.

Figure 8 shows a graph of the Raman spectroscopy results in the 100 to 1100 cm^{-1} wavelength range, where the peaks at 210, 302 and 568 cm^{-1} observed in the spectrum are related to transverse acoustic (TA), longitudinal acoustic (LA) and transverse optical (TO) respectively and refer to the first-order Raman vibration characteristic of TiN thin films.

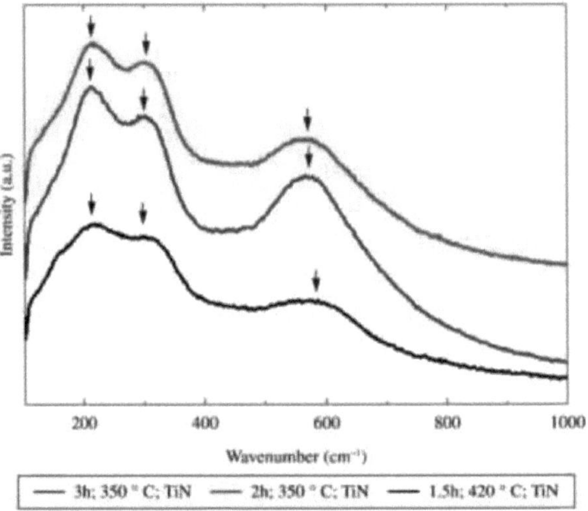

Figure 8 - Raman spectra of samples deposited using a titanium double cage at temperatures of 350°C and 420°C and times of 1.5h, 2h and 3h. (SOUSA et al., 2015)

Another way of demonstrating that TiN has been deposited by double cages Sousa et al. (2015) uses XRD and exposes the patterns of the x-ray diffraction peaks characteristic of titanium nitride films according to figure 9.

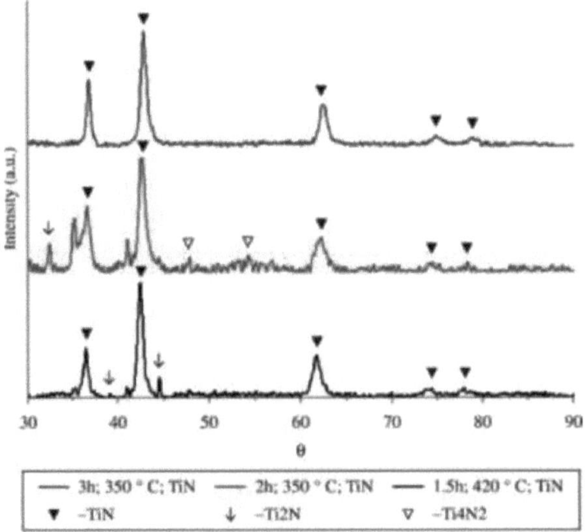

Figure 9 - Shallow X-ray diffractogram of the glass samples deposited using two titanium cages at 350 °C and

420 °C and 1.5, 2 and 3 hours of treatment.

The EDS spectrum, figure 10, serves to reinforce the two other readings in the article by Sousa et al. (2015), which shows the chemical composition of the films deposited on the glass samples. In these spectra, the percentage of titanium on all the deposited surfaces is highlighted.

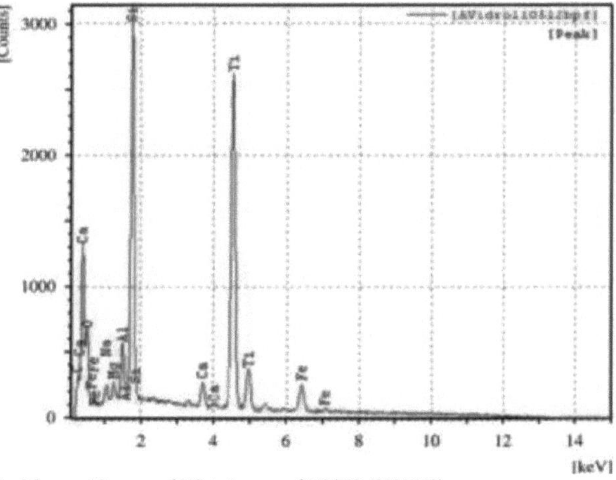

cathodic at 350 °C - 3 hours, 12 sccm of H2 + 4 sccm of N2/H2 (80% N2)

Daudt (2012) aimed to study the deposition parameters and how the different configurations of cathode cages influence the species present in the plasma. Firstly, during the treatment, the author used Optical Emission Spectroscopy (OES).

Among the graphs present in the analysis, two types stand out, showing the values of the luminescent intensity peaks associated with the species present in the plasma as a function of the H2 flow for different cages as shown in figure 11 and the other type of graph shows the values of the luminescent intensity peaks associated with the species present in the plasma as a function of the treatment time for different cages as shown in figures 12 and 13.

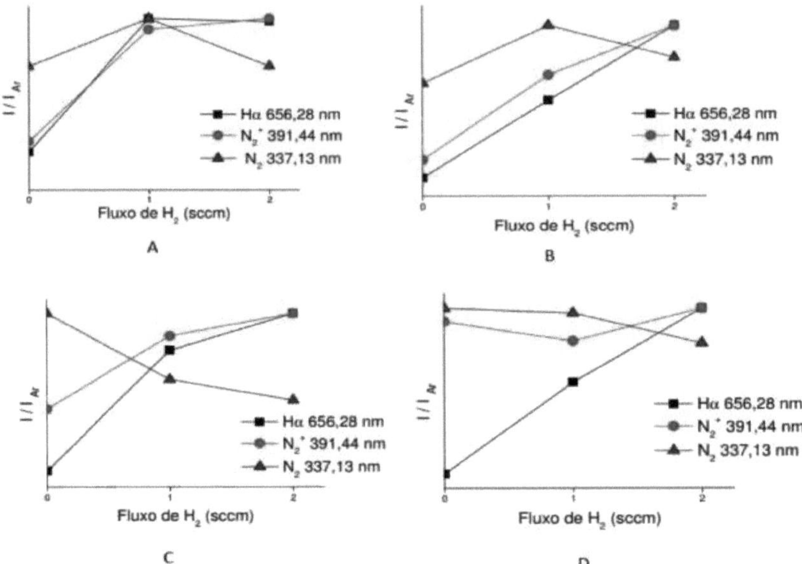

Figure 11 - Profile of the light intensities associated with the species present in the plasma as a function of the addition of H2 gas to the plasma mixture during the nitriding process using a cathodic cage. For cages with (A) 1 series of side holes, 4 holes in the lid with diameters of 8mm; (B) 1 series of side holes, 8 holes in the lid with 8mm diameters; (C) 2 series of side holes, 4 holes in the lid with 8mm diameters; (D) 2 series of side holes, 4 holes in the lid with 8mm diameters.

From the different peak values of the specimens' light intensity for each cathode cage configuration, there may be a correlation between the cage and its configuration and its influence, but these peak changes may also be the result of varying currents during treatment and the presence or absence of hollow cathodes. It is therefore practically impossible to say whether there is an influence or not using OES alone.

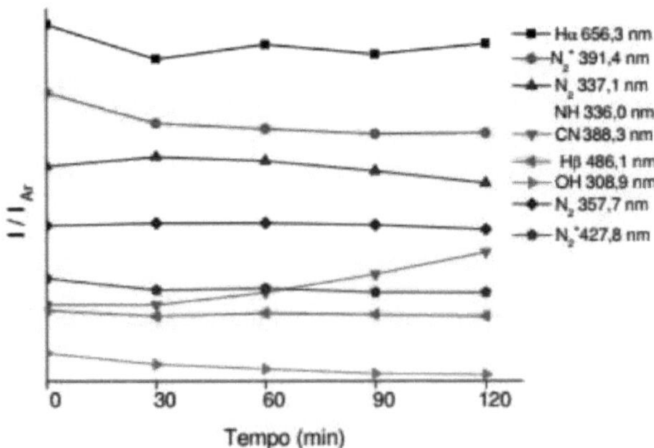

Figura 12 - Luminescent intensity profile associated with the species in the plasma as a function of process time for a cage with 1 series of side holes, 8 holes in the lid with a diameter of 8 mm in a plasma atmosphere with 1 sccm of H_2.

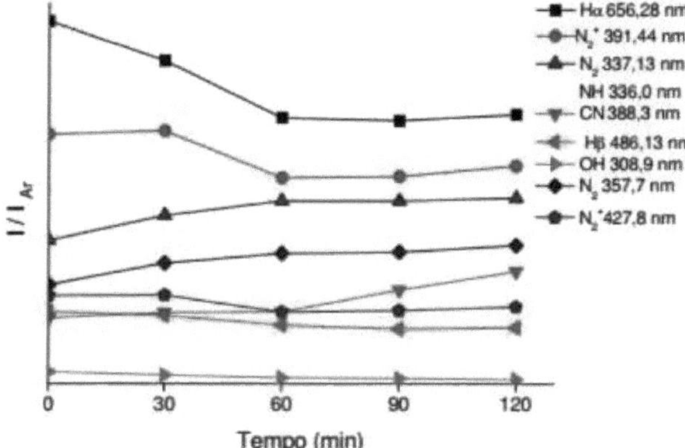

Figura 13 - Luminescent intensity profile associated with the species in the plasma as a function of time process for a cage with 1 series of side holes, 8 holes in the lid with a diameter of 12 mm in a plasma atmosphere with 1 sccm of H_2.

Figures 12 and 13 show the different behaviour of the plasma species with different cages. In both, the OH peak decreases due to the efficiency of the plasma reactor cleaning, while the CN peak comes from the reactor sealing rubber which binds to the reactant nitrogen. In the first image, there are practically no changes in illumination intensity for other species. In the second image we see a decrease in the intensity of the N_2^+, Ha and Hp peaks, and an increase in the relative intensity of the N2 and NH peaks. This probable decrease in the H and N_2^+ species may be related to the formation of NH.

To check for TiN films on the samples from the different treatments, Daudt (2012) used X-ray diffraction with a shallow angle of incidence. There are characteristic peaks in all the graphs, some with lower or higher incidence and they may be related to the amount of TiN deposited on the sample, i.e. the thickness of the film. As shown in Figures 14 and 15.

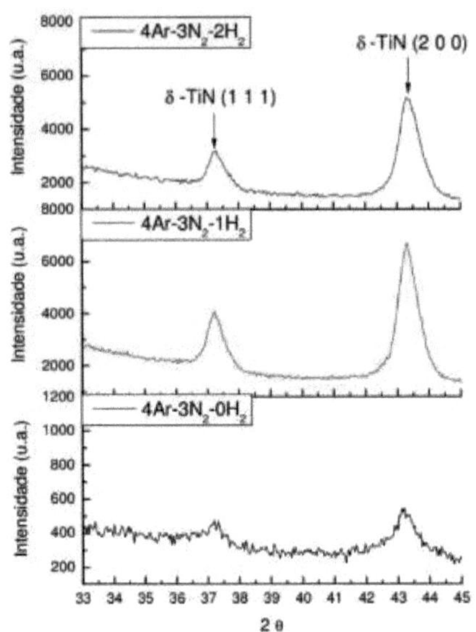

Figura 14 - X-ray diffractogram with a shallow angle of incidence of 0.5° for the deposited films

Figura 15 with the cage with 1 series of side holes, 8 holes in the lid with a diameter of 12 mm for plasma atmospheres as shown in the graph.

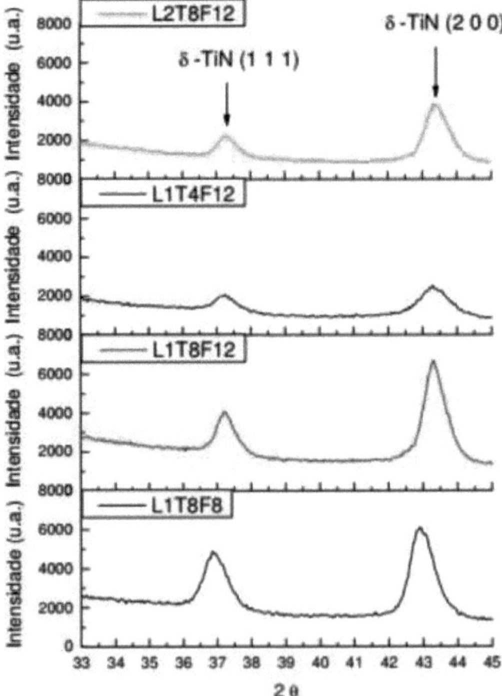

Figura 16 - X-ray diffractogram at a shallow angle of incidence of 0.5° for films deposited with different cages for $4Ar\text{-}3N_2\text{-}1H_2$ plasma atmospheres

It is noticeable that the peaks differ in intensity depending on the number of holes in the lid and the percentage of hydrogen gas flow.

Figures 16, 17, 18, 19 and 20 show the atomic force microscopy readings by Daudt (2012).

Figura 17 - AFM topographic image of the surface of the film deposited with the L1T8F12 cage in the $4Ar\text{-}3N_2\text{-}0H_2$ plasma atmosphere. Image in (a) 3D and (b) 2D.

Figura 18 - AFM topographic image of the surface of the film deposited with the L1T8F12 cage in the $4AR\text{-}3N_2\text{-}1H_2$ plasma atmosphere. Image in (a) 3D and (b) 2D

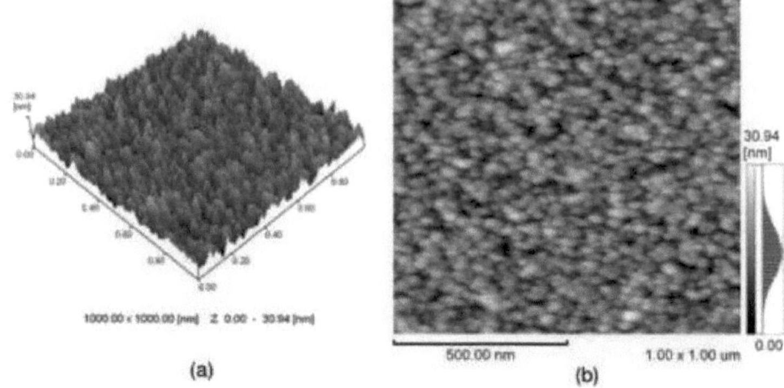

Figura 19 - AFM topographic image of the surface of the film deposited with the L1T8F12 cage in the 4Ar-3N_2-2H_2 plasma atmosphere. Image in (a) 3D and (b) 2D.

Figure 19 - Topographic image obtained by AFM of the surface of the film deposited with the L1T8F8 cage in the 4Ar-3N_2-1H_2 plasma atmosphere. Image in (a) 3D and (b) 2D.

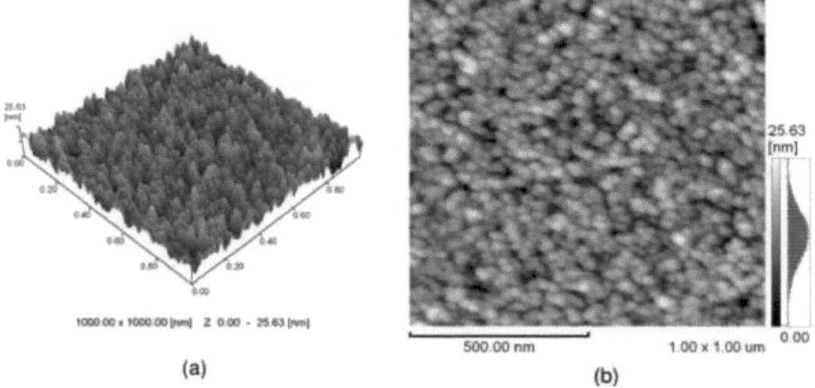

Figure 20 - AFM topographic image of the surface of the film deposited with the L1T4F12 cage in the plasma atmosphere of $4Ar\text{-}3N_2\text{-}1H_2$. Image in (a) 3D and (b) 2D.

When comparing figures 16, 17 and 18, according to Daudt (2012), for which the variable was the hydrogen flow during deposition, it was noted that without the presence of hydrogen (figure 3.8) the film has an amorphous appearance, characterised by a disorganised growth of particles. As the hydrogen flow increased, a more organised growth of the film was observed, with a slight increase in the size of the particles and a small decrease in roughness.

When comparing figures 17, 19 and 20, according to Daudt (2012), for which the variable was the cage used during deposition, the smaller the hole diameter (figure 19), the larger the particles, which shows a higher growth rate of the grains to the detriment of the nucleation rate, there was also a higher roughness parameter, caused by the larger size of the particles.

The optical microscopy images by Daudt (2012) are shown in figures 21 and 22.

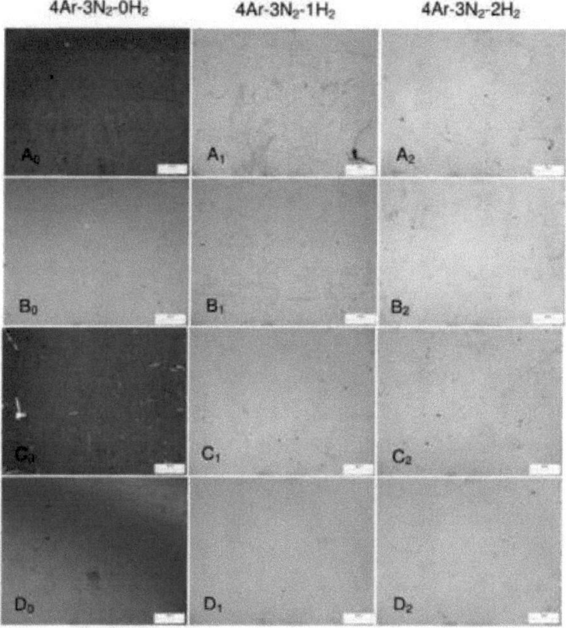

Figure 21 - Micrographs obtained at 50x magnification for the films deposited under different process conditions, with index 0 referring to the $4Ar\text{-}3N_2\text{-}0H_2$ plasma atmosphere, 1 referring to the $4Ar\text{-}3N_2\text{-}1H_2$ plasma atmosphere and 2 referring to the $4Ar\text{-}3N_2\text{-}2H_2$ plasma atmosphere for cages (A) L1T4F8; (B) L1T8F8; (C) L2T4F8; (D) L2T8F8.

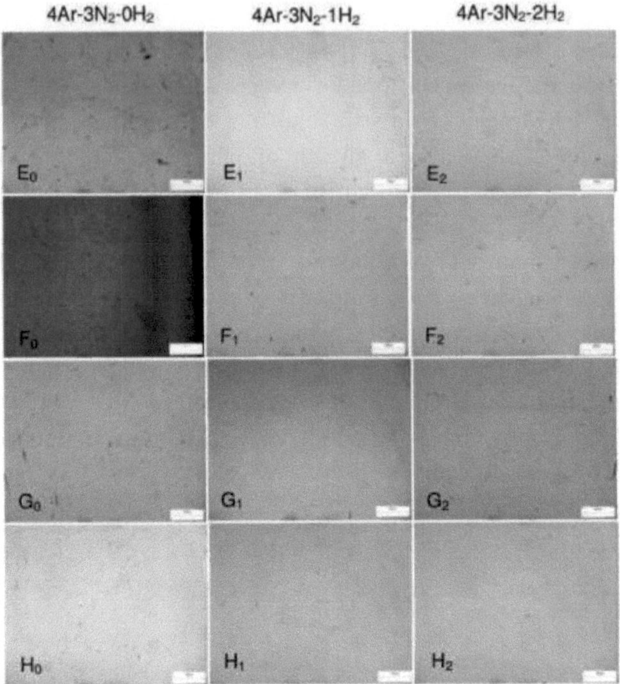

Figure 22 - Micrographs obtained at 50x magnification for the films deposited under different process conditions, with index 0 referring to the 4Ar-3N2-0H2 plasma atmosphere, 1 referring to the 4Ar- 3N2-1H2 plasma atmosphere and 2 referring to the 4Ar-3N2-2H2 plasma atmosphere for cages (E) L1T4F12; (F) L1T8F12; (G) L2T4F12; (H) L2T8F12.

Visually, the samples didn't change much, only those in column 0, showing that the lack of hydrogen leads to inhomogeneous deposition.

The optical properties of Daudt (2012) were studied by spectrophotometry using the results of transmittance and total reflectance and by ellipsometry according to figures 23, 24, 25 and 26.

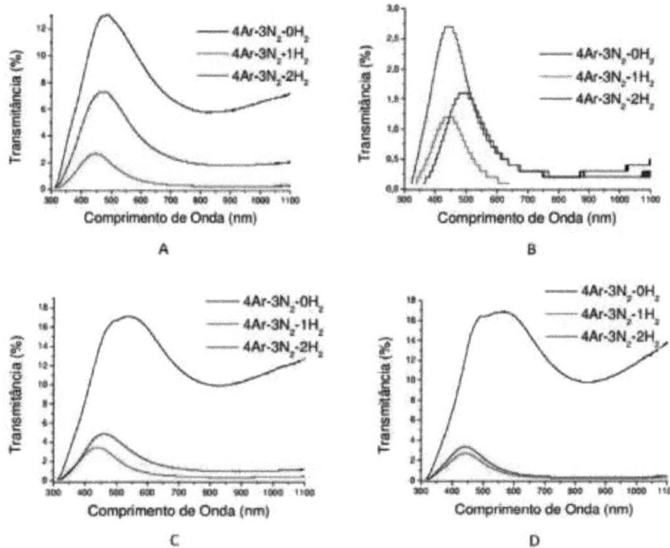

Figura 23 - Transmittance profile for the films deposited in the different plasma atmospheres. For cage (A) L1T4F8; (B) L1T8F8; (C) L2T4F8; (D) L2T8F8.

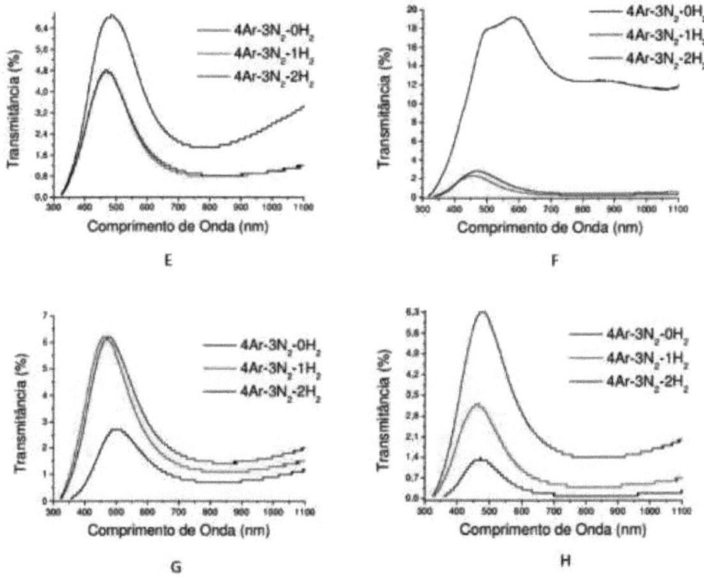

Figura 24 - Transmittance profile for the films deposited in the different plasma atmospheres. For cage (E) L1T4F12; (F) L1T8F12; (G) L2T4F12; (H) L2T8F12.

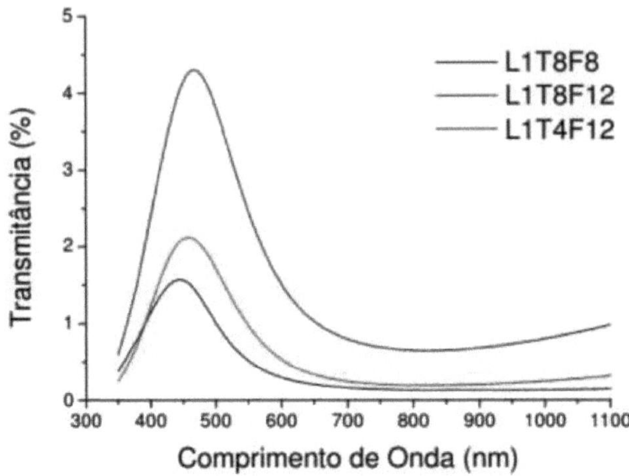

Figura 25 - Transmittance profile for the films deposited for the different cages in the 4Ar-3N2-1H2 plasma atmosphere.

These results show that the transmittance decreases as the hole diameter decreases for all the cage configurations studied, which is related to the increase in film thickness. This increase is probably due to the increase in the number of holes, since with a smaller hole diameter there are more holes in the cage.

The reflectance results are shown in figure 26 by Daudt (2012).

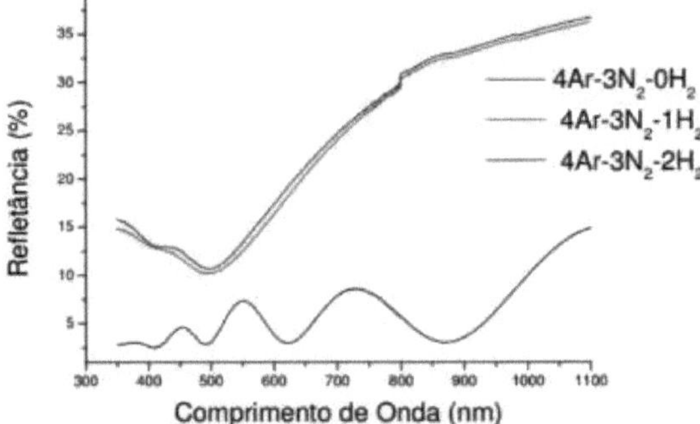

Figure 26 - Reflectance profile for sample obtained with L1T8F12 cage.

According to Daudt (2012), the reflectance results confirm the transmittance results.

Table 5 prepared by Daudt (2012) shows the thickness values and composition for different samples analysed by ellipsometry.

Table 5 - Thickness values (ϕ) and composition

Gaiola	Atmosfera	φ (nm)	TiN (%)	Titânio (%)	Porosidade (%)
$L_1T_8F_{12}$	4Ar-3N_2-2H_2	21,4 ± 5,3	40,7	16,9	42,4
$L_1T_8F_{12}$	4Ar-3N_2-1H_2	21,1 ± 5,2	39,9	15,9	44,1
$L_1T_8F_8$	4Ar-3N_2-1H_2	44,5 ± 7,7	35,0	15,0	50,0
$L_1T_4F_{12}$	4Ar-3N_2-1H_2	21,1 ± 5,4	38,3	17,8	43,8

Source: (DAUDT, 2012)

The results in table 5 show that when comparing the different cage configurations, as expected, the film with the lowest transmittance (L1T8F8 cage) has the greatest thickness. for the other films analysed the thickness is practically the same, with variations within the error.

Silva et al. (2017) referenced the glass samples according to the thickness of the lid and the test number as shown in figure 27. The authors aimed to check the deposition rates and optical properties.

Figure 27 - Photograph of the pairs of films obtained on the glass substrate. The names E1, E2, E7 refer to the test and the names G1 and G10 refer to the test.

In figure 28, Silva et al. (2017) shows the X-ray diffractograms for three pairs of films

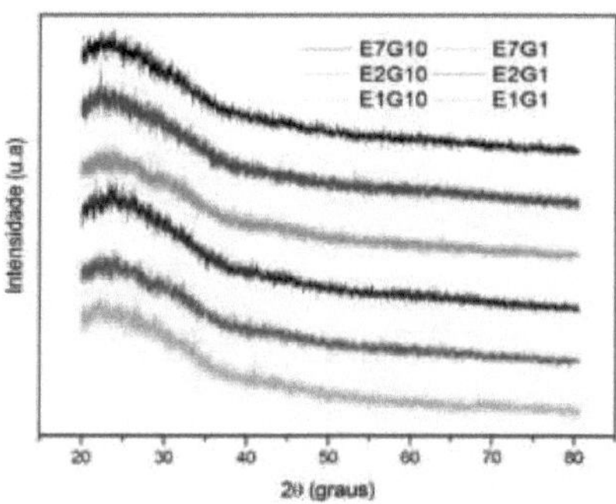

Figure 28- Diffractograms of the thin films of samples E1G1, E1G10, E7G1 and E7G10

The graphs do not show any characteristic TiN deposition peaks like the same analyses by Daudt (2012) or Sousa et al. (2015). In comparison with other authors, the authors justify the low temperature and short duration as possible causes of amorphous films.

The reflectograms of the films are summarised in Table 6. To fit the theoretical curves using Philips Wingixa software, two models were tested. The first model was a film with a single layer on the substrate and the second model, a film composed of two layers, the first thinner surface layer and a second thicker layer on the substrate. The density of the glass was kept fixed at 2.2 g/cm^3 . The other variables were left free.

Of the two models, the best fits were obtained using the two-layer model and the results of the fits are shown in Table 6. The second layer could possibly be the result of surface oxidation of the TiN film causing (producing) a surface layer, possibly of TiON.

Table 6 - Parameters used to adjust the theoretical RRX curves.

Sample	Thickness (nm)			Roughness (nm)			Density (g/cm)^3			X2
	TiN	TiON	Total	glass	TiN	TiON	glass	TiN	TiON	
E1G1	13,4	2,3	15,7	2,0	2,5	0,8	2,2	3,8	2,9	0,08
E1G10	35.9	2,6	38,5	2.0	4.7	0.7	2.2	3.7	1.3	0.08
E2G1	12,8	2,2	15,0	2,0	1,6	0,9	2,2	3,7	2,9	0.06
E2G10	36,9	3,2	40,1	2,1	2,3	0,6	2,2	3,6	1,4	0,03
E7G1	28,2	3,3	31,5	1,9	1,8	0,7	2,2	4,1	3,1	0,06
E7G10	122,0	3,0	125,0	2,0	4,8	0,5	2,2	5,0	1,4	0.09

Source: (SILVA et al., 2017)

It can be seen that regardless of the flows used, a thicker lid produces an increase in the thickness of the films and, consequently, a greater surface roughness of the films, since the higher the deposition rate, the greater the roughness. In addition, higher energy particles produced in the denser plasma regions of the thicker lid may also contribute to greater roughness in these films.

Figure 29 shows the transmittance curves for the six films studied. The E7G10 film showed the lowest transmittance, which is probably due to its greater thickness.

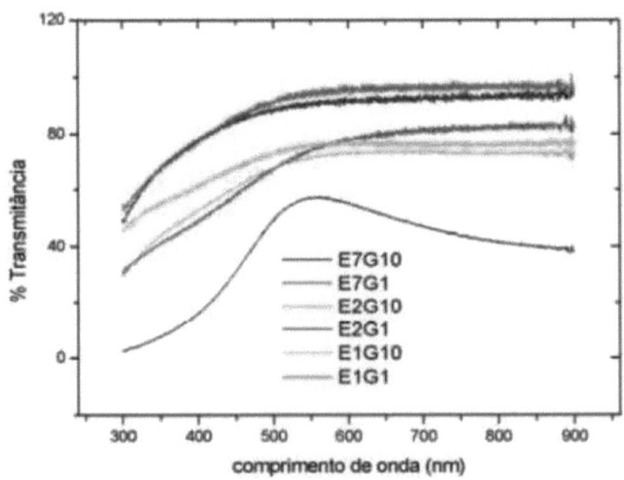

Figure 29 - Transmittance of films deposited on glass substrate.

This film shows a sharper drop in transmittance in the red region than the others. This may indicate that this film is the one that came closest to the stoichiometric titanium nitride, since it has low infrared transmittance.

The EDS scan of the surface of each film in a central square region with a side of 100 pm identified the elements present in the substrate/film and their respective quantities in weight per cent. These data are shown in Table 7. Looking at the titanium element column, we can see a greater quantity in the films that were deposited using the 10 mm thick lid cage, as had already been seen in the reflectometry data.

Table 7 - Percentage by weight of the elements contained in the film/substrate

ELEMENTS

SAMPLE	O	Yes	In	Ca	Mg	Al	Ti	N	K	total
E1G1	52,4	31,4	9,4	3,5	7 7	0,7	0,3	0,0	0,0	99,9
E1G10	52,0	31,9	8,1	3,4	1,9	1.1	0,8	1,1	0,2	100,5
E2G1	52,4	31,0	9,6	3,6	2,2	0,6	0,3	0,3	0,0	100
E2G10	51,6	31,8	9,2	3,7	2,3	0,0	0,8	0,4	0,0	99,8
E7G1	50,7	30,9	9,7	3,8	2,1	0,6	0,7	0,3	0,2	99
E7G10	43,9	32.5	8,7	3,9	2.1	0,6	3,0	5.1	0,0	99,8

Source: (SILVA et al., 2017)

The results for the E2 and E1 films were the same, 0.8 when using the G10 cage and 0.3 when using the G1 cage. In other words, by this measure, the increase in the amount of titanium in the film when using the G10 cage and a pressure of 1.0 Torr was the same, regardless of the two gas concentrations used. However, when using a lower pressure of 0.5 Torr and an intermediate gas concentration, the increase in the amount of titanium deposited was much greater when using the G10 cage.

4 FINAL CONSIDERATIONS

Titanium nitride films can be obtained by plasma nitriding in a titanium cathode cage and have many uses in industry. According to the references and authors studied on the deposition of thin titanium nitride films by plasma nitriding on glass samples using a cathodic cage, they suggest that they are formed and deposited on the glass, depending on the proximity of the cage, the temperature and the treatment time. They also suggest that the thickness of the film increases as the thickness of the cage cover increases and as the diameter of the hole in the cage decreases, thereby reducing its transmittance, proving that it can also be used as a solar control system.

REFERENCES

DAUDT, N. DE F. **Influence of process parameters on titanium nitride deposition by cathodic cage plasma**. Dissertation (Master's Degree in Processing Materials from Dust; PoEmeros and Composites; Processing Materials from Dust)-Natal: Federal University of Rio Grande do Norte, 2012.

FERNANDES, F. DE M. et al. DEPOSITION OF THIN COPPER FILMS BY CATHODIC GAIOL: ANALYSIS OF PLASMA CONFINEMENT AS A FUNCTION OF INCREASING CATHODIC GAIOL COVER THICKNESS. **Technology in Metallurgy, Materials and Mining**, v. 15, n. 3, p. 296-302, 2018.

HOWATSON, A. M. **An Introduction to gas discharges**. 2d ed. Oxford; New York: Pergamon Press, 1976.

SILVA, S. S. DA et al. Plasma deposition of titanium nitride thin films with hollow cathode length effect in cathode cage. **Materia (Rio de Janeiro)**, v. 22, n. 3, 10 Aug. 2017.

SOUSA, R. R. M. DE. **Cathodic cage plasma nitriding: investigation of the mechanism and comparative study with continuous tension plasma nitriding**. Thesis (Doctorate in Processing Materials from Dust; PoEmeros and Composites; Processing Materials from Dust) - Natal: Federal University of Rio Grande do Norte, 2007.

SOUSA, R. R. M. DE et al. Thin Tin and Tio2 Film Deposition in Glass Samples by Cathodic Cage. Materials Research, v. 18, n. 2, p. 347-352, Apr. 2015.

YUSTE, M. et al. Improving the visible transmittance of low-e titanium nitride based coatings for solar thermal applications. **Applied Surface Science**, v. 258, n. 5, p. 1784-1788, dec. 2011.

ZHENG, P. et al. Study of titanium nitride for low-e coating application. **Chinese Science Bulletin**, v. 52, n. 13, p. 1860-1863, jul. 2007.

ALVES JUNIOR, Clodomiro. **Plasma nitriding: fundamentals and applications**. Natal: Edufrn, 2001. Available at: https://repositorio.ufrn.br/jspui/handle/1/11807. Accessed on: 21 Apr. 2021.

RIBEIRO, Kleber Jose Barros. **Plasma nitriding with cathodic cage: characterisation and performance evaluation of the nitrided layer on cutting knives**. 2007. 92 f. Dissertation (Master's Degree) - Course in Processing Materials from Dust; Polymers and Composites; Processing Materials from Dust, Federal University of Rio Grande do Norte, Natal, 2007. Available at: https://repositorio.ufrn.br/handle/123456789/12854. Accessed on: 21 Apr. 2021.

PINEDO, C. E. **Surface treatments for tribological applications**. Caderno tecnico Metalurgia & Materiais. April 2004.

MEDANHA, A; GOLDENSTEIN, H; PINEDO, C.E. **Heat Treatments of AISI D2 Tool Auger for Forming and Cutting**. Available at: . Accessed on: 01 August 2011.

KOCH, H., FRIEDRICH, L.J., HINKEL, V., *et al.,* **Hollow Cathode Discharge Sputtering Device for Uniform Large Area Thin Film Deposition**, *J. Vac. Sci. Technol. A*, v.9, n.4, pp. 2374-2377, 1991.

CHAPTER 3

SUMMARY

This paper presents a system for inspecting welded joints in industrial production lines for pipes and profiles. The system is based on computer vision techniques and is capable of detecting non-conformities in real time. The results showed that, out of 300 tests carried out, the system had an error rate of 4.67%, indicating that it can identify anomalies in most cases. However, there were 14 type two errors, showing that there is still room for improvement in reducing these errors. In addition, the data also showed that the majority of errors occurred in galvanised material, followed by cold-rolled and hot-rolled material. Despite achieving an accuracy rate of 95.33 per cent, the system still has a vulnerability, especially in lighter-toned materials. This is due to the greater reflectivity of the material, which can cause difficulties in recognition, leading to failure. The system's potential for use in difficult-to-access environments, in order to contribute to the safety of operational personnel, is also highlighted. Future research could investigate non-dye-based detection methods, which could reduce operating costs, since the dyeing process involves a cost in time and money. Therefore, the system presented is a promising solution for detecting welded joints on industrial production lines, and opens up scope for further research.

Keywords: welded joints; inspection; pipes; profiles; automation.

1INTRODUCTION

Visual inspection is a technique widely used in the assessment of components and structures to detect surface and subsurface anomalies that could affect the performance or use of the inspected object. This analysis consists of inspecting the object using the human eye with or without the aid of magnifying glasses to detect anomalies. It is considered the most common and fundamental method for detecting defects and discontinuities (CHAKRABORTY, 2018; KUMAR, 2020).

The field of computer vision emerged with the aim of replicating human vision and automating tasks that require visual analysis and understanding. This area uses technologies to acquire and digitise images or videos of objects or scenes using cameras, and to extract information from these images using image processing techniques. The application of these concepts and methods in industry is known as computer vision, encompassing any use of vision techniques to perform industrial tasks. Automated visual inspection is the most developed area of computer vision, used to check the quality of manufactured products (SZKILNYK, 2012; LEVINE, 1985).

Increased competitiveness in industry has encouraged the development of strategies to optimise processes and value chains. As a result, the use of computer vision in industry has been growing in recent years, covering a variety of fields. This is because computer vision methods are used as the basis for the development of widely used technologies, such as simultaneous visual mapping and localisation, object tracking and non-conformity identification, among others. Therefore, this work proposes a system that uses computer vision tools to increase the efficiency and reliability of the visual inspection process at an industrial site (MRUGALSKA; WYRWICKA, 2017; KAKANI et al, 2020).

Welding is an inherent process in the manufacture of steel tubes and profiles for machine fuelling. However, under different circumstances, it may be necessary to remove the welded products. In the case of profiles, they are rejected because they are structural elements. As for pipes, the decision whether or not to remove the welded products is left up to the customer, in order to meet their specific requirements.

In order to accurately identify and remove welded products, the operator needs to be vigilant at all times, especially on lines that operate at high speeds. Identifying welded products on such lines can become a challenging activity. Fatigue and the need for agility can compromise the operator's judgement. With this in mind, the project aims to propose an alternative system for automatically identifying welded joints using computer vision.

The identification of welded joints in tubes and profiles is a recurring process in the day-to-day life of the manufacturing industry. In Brazil today, it is common for many of these processes to be carried out manually, which can lead to inaccuracies and human error. In addition, the manual process sometimes puts the operator in risky situations, especially on lines operating at high speeds. Therefore, there is a need to develop efficient, automated alternatives for identifying welded joints in pipes and profiles (YAMADA; MARTINS, 2019).

Computer vision is a technology that is gaining popularity in many industrial sectors. It uses image processing algorithms to analyse and interpret digital images, making it possible to detect objects and identify specific features in an image. In particular, the application of computer vision to the identification of welded joints in pipes and profiles can significantly improve the safety, accuracy and efficiency of the identification process. Furthermore, the technology can be implemented in an automated system, reducing the time and cost of the process (MARTINS; JUNIOR, 2011).

Based on this context, the aim of this project is to propose an alternative for identifying welded joints in pipes and profiles using computer vision. The project aims to explore the application of computer vision in the detection of welded joints in pipes and profiles, identifying the technical and operational requirements for implementing the technology. By developing an automated image processing system, it is hoped to improve the quality and efficiency of the welded joint identification process, reducing the costs associated with the manual process and increasing the competitiveness of companies operating in this market.

Given the problems identified, hypotheses were drawn up with the aim of understanding them, solving them and proving or disproving the assumptions. Below are the hypotheses formulated:

- Carrying out the activity independently of human action can increase the reliability of the process.
- Eliminating human intervention during the process can provide greater safety for the operating team.
- Reducing the number of faults in the process can convert the time lost to rework into an increase in production.

1.1 GENERAL OBJECTIVE

Proposing a system for identifying welded joints in pipes and profiles using computer vision.

1.2 SPECIFIC OBJECTIVES

- Develop an image processing algorithm suitable for identifying welded joints;
- Test and evaluate the effectiveness of the image processing system developed through simulation;
- Compare the results obtained with the proposed image processing system with current methods for identifying welded joints;

2 THEORETICAL FRAMEWORK

2.1 INDUSTRIAL AUTOMATION

Industrial automation is a process in which operations are carried out using machines and equipment that are controlled by advanced software and intelligent systems, making the production process more efficient and less dependent on human labour. From this definition, it is clear that industrial automation is a concept that involves the use of advanced technologies to improve productivity, product quality and the competitiveness of companies. Industry 4.0, which is one of the most recent industrial automation concepts, has a major impact and a wide range of changes in manufacturing processes, results and business models (BAHRIN et al, 2016).

Industrial automation is important because it enables customised mass production, increasing productivity, flexibility and production speed and improving product quality. In addition, industrial automation also encourages innovation, since prototypes or new products can be produced quickly without the need to reconfigure machines or set up new production lines (BAHRIN et al, 2016). Integrating product development with digital and physical production has been associated with major improvements in product quality and a significant reduction in error rates. In addition, collecting and analysing data using "big data" techniques makes it possible to identify and solve small problems in real time, increasing quality and reducing costs (AZAMFIREI; PSAROMMATIS; LAGROSEN, 2023).

This requires companies to balance economic, social and environmental factors. Furthermore, quality management is a critical aspect of manufacturing, as poor quality can have negative financial, environmental and social impacts. Automated processes are necessary to achieve the required levels of flexibility, speed and precision needed to meet customer demands (MRUGALSKA; WYRWICKA, 2017).

2.2 COMPUTER VISION

Computer Vision is a field of Artificial Intelligence (AI) that uses algorithms to analyse images and videos and extract relevant information from them (SILVA et al, 2018). The application of Computer Vision in industrial inspections is a way of guaranteeing the quality of products and processes, as well as being a solution for tasks that would be difficult or impossible for a human inspector to perform. The use of Computer Vision systems enables the detection of defects in handles with high precision and speed, as well as improving the efficiency of the production process (AZAMFIREI; PSAROMMATIS; LAGROSEN, 2023).

The Computer Vision System consists of a combination of hardware and software capable of performing visual inspection tasks. The hardware consists of cameras, lighting and positioning control systems to capture precise, repeatable images of the object under inspection. The software is responsible for analysing the images and identifying defects or variations in relation to a reference model (LABUDZKI; LEGUTKO; RAOS, 2014). For the system to work properly, it is necessary to train the software with reference images and adjust the lighting and camera parameters for each specific case (SILVA et al, 2018).

The applications of computer vision in industrial inspections are diverse and can be used in different sectors such as the automotive, electronics, food and pharmaceutical industries, among others. In the automotive industry, Computer Vision has been widely used in tasks such as paint inspection, faulty handle detection, component recognition and quality control in general. The use of AI techniques, such as neural networks and machine learning, has enabled the continuous improvement of Computer Vision systems, making them increasingly accurate and efficient for the inspection of industrial products (SILVA et al, 2018).

2.3 RASPBERRY PI

The Raspberry Pi is a mini-computer designed with the aim of helping young people learn the basics of computer programming at an affordable cost. Since its creation in 2009 by researchers at the University of Cambridge, the Raspberry Pi has gained popularity all over the world. The device was designed to increase the number of students opting for computer science by providing an accessible and versatile learning platform (UPTON; HALFACREE, 2016).

One of the biggest advantages of the Raspberry Pi is its wide availability on the market and low cost, making it accessible to companies of all sizes. In addition, the Raspberry Pi supports a variety of programming languages, including Python, C, C++, BASIC, Perl and Ruby, making it a flexible and customisable platform to meet the needs of any company (BHATTACHARYA; BHATTACHARYA, 2019).

Another great advantage is the support for a large number of peripherals, including 26 GPIO (General Purpose Input/Output) pins, allowing various external devices to be connected. This makes the Raspberry Pi an ideal tool for industrial automation projects, process control, access control, security and much more (FONSECA; MACEDO, 2018).

Despite some limitations, such as the lack of internal storage, the Raspberry Pi is an attractive option for companies looking for a low-cost solution for industrial automation and IoT (Internet of Things) projects. In short, the Raspberry Pi is a versatile and affordable tool that offers a multitude of possibilities for companies in all sectors. Whether for industrial automation, process control or IoT projects, the Raspberry Pi is a low-cost and highly customisable option that can meet the needs of industry (MATHE et al, 2022).

2.4 OPENCV

OpenCV is a programming library widely used in the creation of computer vision applications, with various tools for image and video processing that make it possible to create applications capable of recognising patterns, tracking objects and performing other tasks that involve analysing visual data (ZHANG et al, 2017). A major advantage of OpenCV is its open source nature, allowing developers to use and modify the code without licensing restrictions.

In industrial environments, computer vision applications based on OpenCV can provide significant benefits. These applications can help automate various processes, such as inspection, sorting and assembly, leading to greater efficiency and cost savings. For example, computer vision-based systems can be used to inspect products and detect defects in real time, eliminating the need for manual inspection and reducing the risk of errors. In addition, computer vision systems can be used to track and sort products on a production line, ensuring that they are delivered to the correct location in a timely manner (BISWAS et al, 2018).

Despite its many advantages, there are also some disadvantages to using OpenCV. One of the main challenges is the complexity of the library, which can make it difficult to use effectively for developers without extensive experience in computer vision. In addition, OpenCV can be processing intensive, requiring significant processing power to analyse large volumes of visual data in real time. However, with the right expertise and hardware resources, these problems can be solved and the benefits of using OpenCV in industrial applications can be realised (SILVA et al, 2020).

OpenCV is a powerful tool that can be used to develop a wide range of computer vision applications in industrial environments. The library's open source nature and comprehensive functionalities make it an attractive choice for developers looking to create automated systems that can analyse and process visual data in real time. Although there are some challenges associated with using OpenCV, such as its complexity and processing requirements, these can be solved with the right expertise and resources (SILVA et al, 2020).

3 METHODOLOGY

3.1 DEFINITION OF NON-CONFORMING PRODUCT

The concept of a non-conforming product is broad and varies according to the context in which it is used. In this work, a non-conforming product is one with a weld bead from the machine refuelling stage.

During the refuelling process, the material to be inserted into the machine must be joined to the material remaining from the previous refuelling by welding. Figure 1 shows this process being carried out.

Figure 30 - Refuelling weld.

Source: author.

In addition, to make it easier to identify the product with the weld seam, it is common practice in the company where the study was carried out for the welded area to be coloured red (Figure 2).

Figure 31 - Sheets with and without paint.

Source: author.

3.2 NON-COMPLIANT PRODUCT SEPARATION PROCESS

The separation stage takes place during baling, for which both the tube-forming and roll-forming machines have a piece of equipment attached for automatic bundling and wrapping. This equipment is manufactured by OMP and can be seen in Figure 3.

Figure 32 - Automatic packaging machine.

Source: author.

However, the identification of non-conforming products is an input that needs to be indicated by the machine operator, which is done by means of a key located on the packaging machine panel, as can be seen in figure 4.

Figure 33 - Non-compliant product removal panel and key.

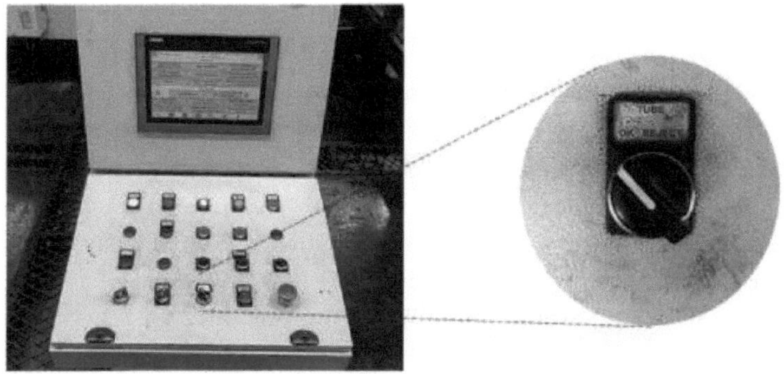

Source: author.

After the switch is actuated, the product is removed by rotating a gripper that redirects the product out of the line, as shown in figure 5.

Figure 34 - Claw for removing non-conforming product.

Source: author.

3.3 DETERMINING THE COMPONENTS

In order to automate the process of identifying non-conforming products, making it independent of human action, a survey of the necessary components was carried out. Considering that the project is proposed as a low-cost automation alternative, priority was given to choosing components in line with this proposal, without compromising the standard of reliability in their function.

Based on the characteristics presented in section 2.3 and the proposal described in the previous paragraph, we chose Raspberry PI, a low-cost computing platform that can be easily integrated with other technologies, making it a versatile option for managing the logical part of the project. To accompany the Raspberry PI, the "camera module 3" was selected, a compact camera with a 12 megapixel sensor for taking images. In addition, other electronic components are also required, such as a 5 V supply for powering the system, a relay for activating the key, a radio frequency module to allow the relay to communicate with the Raspberry PI and a case for mounting and protecting the components.

3.4 ARRANGEMENT OF COMPONENTS

The arrangement of the components is a crucial factor in guaranteeing the system's proper performance. To this end, the need to protect the system components from external factors such as vibration, mechanical shocks and the possibility of being hit by soluble oil was considered. In this sense, the best place to install the camera is on the machine's scanning table, since the machine is not at risk of mechanical shock or being hit by a jet of soluble oil, and the location provides a larger free area for installation and an unobstructed view of the product. Figure 6 shows a perspective view of the camera.

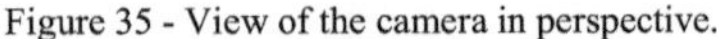
Figure 35 - View of the camera in perspective.

Source: author.

Figure 7 shows the layout of the components and a simplified version of their workflow.

Figure 36 - Simplified version of the component layout.

Source: author.

3.5 PYTHON CODE DEVELOPMENT

In order to reduce the costs associated with carrying out the tests, two similar codes were developed, one with its syntax geared towards running on a

Raspberry Pi and the other for a standard computer. This strategy was adopted with the aim of enabling the tests to be carried out using components that were already available, thus avoiding the need to invest in new equipment. In this context, this work was conducted by analysing the system's performance using a laptop and the camera of a smartphone. Both codes are available (Appendix A and B) with comments to explain how they work.

For comparison purposes, the smartphone camera is capable of recording full HD videos at 30 frames per second, while the camera proposed for the project is capable of recording full HD videos at 50 frames per second. Although the results achieved in the tests do not have the same level of performance expected for the proposed system, it is a very similar condition, allowing us to infer that the results obtained can be reproduced using the equipment indicated in the previous section.

In order to enable the system to identify the products with the weld bead, a code was devised that could identify the difference in colour in the welded area compared to the rest of the product. Initially, it was determined that if the system identified dark colours, the product should be classified as non-compliant and discarded. Figure 2 illustrates this difference in shades in the welded area compared to the galvanised auger sheet. However, during tests with other types of steel, such as hot-rolled steel, which has a darker colour due to the rolling process, false positives were observed in the identification of the weld bead.

For this reason, it was decided to adopt the strategy already used by the operational team of painting the welded area (Figure 2) and reprogramming the code to identify the colour red, in order to eliminate the presence of false positives. With this new strategy, it was possible to increase the reliability of the results and make the system more efficient at detecting the weld bead on products.

3.6 VALIDATION OF RESULTS

The work was carried out using tests in a controlled environment to simulate different working conditions. To do this, tests were carried out on three types of raw material: galvanised (A), cold-rolled (B) and hot-rolled (C), as can be seen in Figure 8. This choice was made because the colour of the product varies depending on the process it has undergone, so the aim was to see if these differences would affect the system's identification process.

Figure 37 - Specimens used.

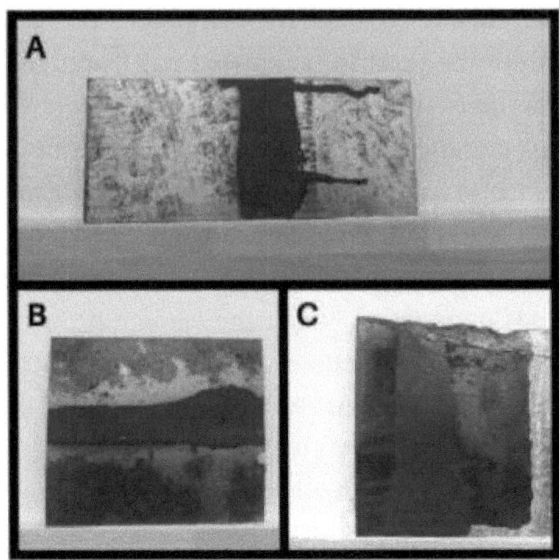

Source: author.

To validate the results obtained, it is important to consider the operating conditions to ensure that the simulation is accurate. In this sense, the main characteristic to be evaluated is the speed of the production line, because the higher the speed, the greater the possibility of the system not being able to detect the moving object. The rollformer and the tube former operate at different speeds, but the tube former operates at a higher speed. Therefore, for the purposes of this analysis, the maximum speed was considered to be
reached by the tube former, since if the system performs satisfactorily in this configuration, it is possible to conclude that it is also capable of operating at lower speeds.

The maximum operating speed of the tube-forming machine in the company where the study was carried out is 145 metres per minute. To check whether the system is capable of identifying a non-compliant product travelling at this speed, a test environment was developed. The approach used consisted of leaving the body in free fall, so that when it passes through the sensor's field of view, the object is already moving at the desired speed. To ensure this, Tracker, a free video modelling and analysis tool, was used. By tracking the positions of the test specimen over the frames (Figure 9), it was possible to construct a graph of the position as a function of time.

Figure 38 - Tracking the positions along the frames.

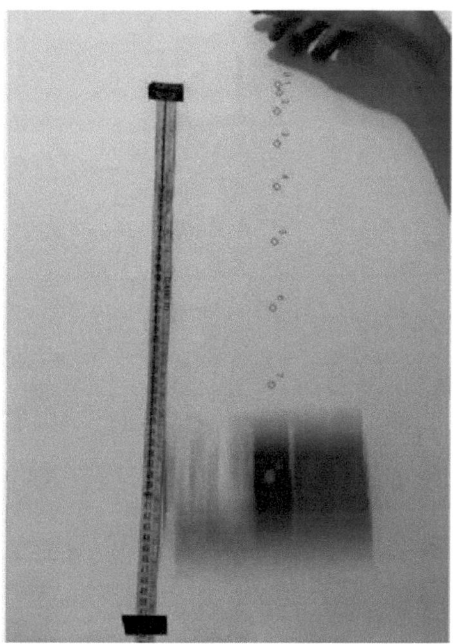

Source: author.

In addition, to check that the results were as expected, the linear regression of the velocity curve as a function of time was carried out. Since the formula governing free fall is v = g *t, where v is velocity, g is gravity and t is time, it was expected that the variance would be low, i.e. R^2 would be very close to one. The structure of the test bench can be seen in Figure 10, where (A) refers to the place where the specimen should be dropped, and (B) is the area where the camera is positioned. Therefore, for the test to be successful, it is enough for the system to be able to identify the welded joint as the object passes through the monitored area.

Figure 10 - Test environment configuration.

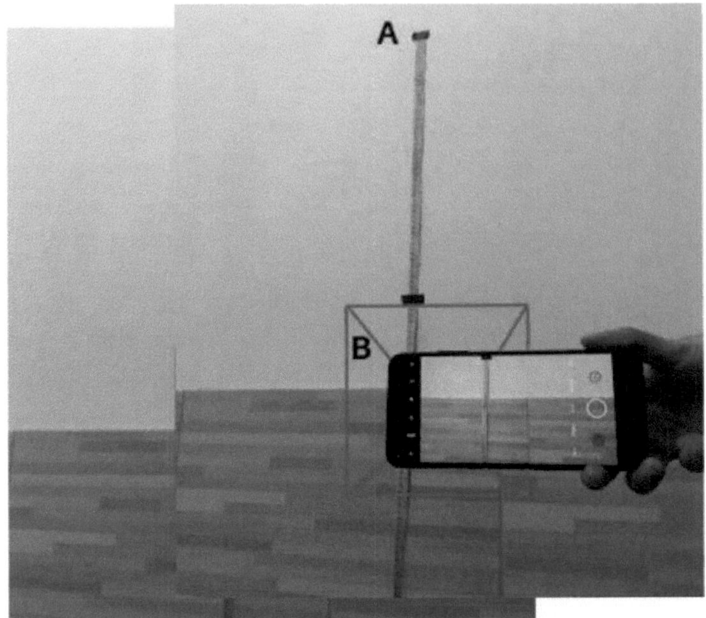

Source: author.

4 RESULTS

Figure 11 shows the velocity versus time graph for the body abandoned in free fall. To confirm that the object reached a speed of 145 m/min (2.4 m/s), a linear regression was performed a priori and the variance was calculated. The result was a coefficient of determination R^2 of 0.9998. Furthermore, it can be seen that the result is in line with what was expected when using the free fall formula. Adopting 2.4 m/s in the equation v = g * t, both methods show that the speed was reached at the instant t = 0.24 seconds, indicating that the data obtained was consistent with the theory.

Figure 11 - Graph of the body's velocity as a function of time.

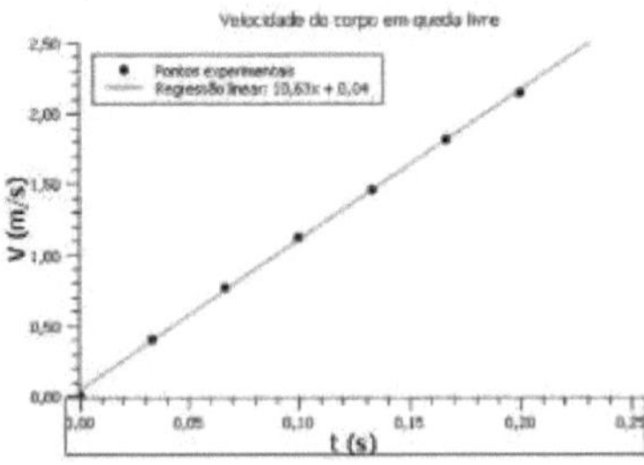

Source: author.

With regard to the classification of results, the following definition was adopted: if the system is able to identify the welded joint, the result will be considered a hit. In the event of a false positive, the result will be classified as a type 1 error, and if the system is unable to identify the welded joint during the test, it will be classified as a type 2 error (STALLARD; MACKENZIE; PETERS, 2018). For greater clarity in identifying the results during the tests, the system was programmed to circumscribe the area in which the weld bead is located. This process takes place in real time and can be seen in Figure 12.

Figure 12 - System recognising the welded joint during one of the tests.

Source: author.

A total of 300 tests were carried out, which were evenly distributed between the three types of auger under study: cold-refined, hot-refined and galvanised. As a result, the system had a hit rate of 95.33 per cent, with only 14 occurrences of type 2 errors detected, while there were no type 1 errors. The distribution of errors between the types of raw material can be seen in Table 1.

Table 8 - Distribution of errors by raw material.

	Galvanised	Hot refined	Cold refined
Type 1 error	0	0	0
Type 2 error	9	1	4

Source: author.

Further analysis revealed that the galvanised steel showed a concentration of flaws, possibly due to its high reflective capacity. To clarify this behaviour, new tests were carried out and, as a result, it was possible to verify the influence of reflection to the detriment of the precise identification of the welded joint by the system. Figure 13 illustrates how the system is affected in this scenario.

Figure 393 - Effect of reflection on the sensor.

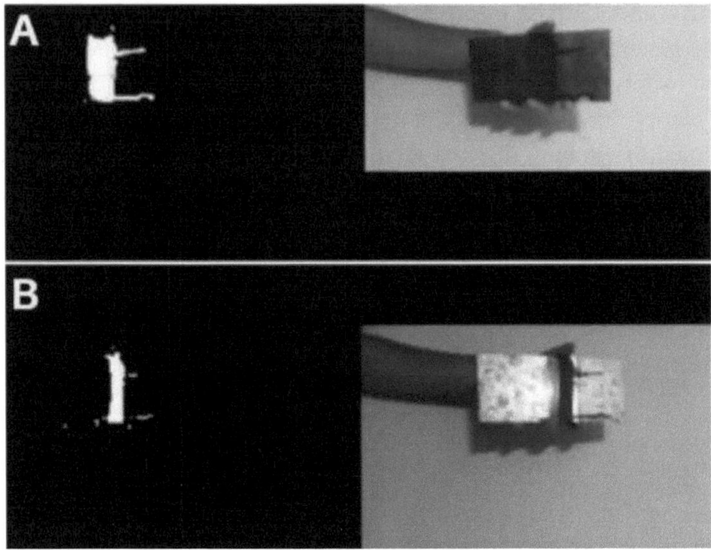

Source: author.

The blank area in the image symbolises the portion captured by the sensor before the application of the soil removal process through erosion and dilation. There is a clear reduction in the area identified when under the influence of the reflection. This reduction is very important since the system was designed to minimise false positives by considering a minimum size for the detected region. In this way, reducing the area could result in an identified area smaller than the minimum size stipulated by the system, which would lead it to classify it as red, thus causing an error. The section of the code responsible for delimiting the minimum size is highlighted in Figure 14.

Figure 404 - Excerpt from the code that determines the minimum area.

```
# Apenas prosseguir, se o raio do contorno possuir um tamanho mínimo.
# Ajustar conforme necessidade.
if radius > 5:
```

Source: author.

Therefore, in order to solve this problem, it is suggested that the value be adjusted according to the specific need, in order to minimise the probability of this type of failure occurring. It should be noted that the value 5 is a dimensionless quantity and therefore must be defined empirically for each application. Furthermore, an alternative solution would be to check, when installing the equipment, whether reflections are likely to occur and to position the camera appropriately to avoid this problem.

5CONCLUSION

5.1 SYSTEM PERFORMANCE EVALUATION

Based on the information presented, it can be concluded that the welded joint inspection system is a promising solution for detecting non-conforming products on industrial tube and profile production lines. The results obtained showed that the system is able to identify most of the specified products, with an accuracy rate of 95.33%. However, there is still room for improvement in reducing type two errors, which occur mainly in materials with a higher reflective capacity.

Furthermore, it is important to emphasise that the system also has the potential to be used in environments that are difficult to access, which can increase the safety of the operational team. Through the use of advanced technologies, such as computer vision, it is possible to automate inspection processes that previously required time and human effort.

5.2 SUGGESTIONS FOR FUTURE WORK

As a suggestion for future work, we recommend investigating methods of detecting welded joints without the need to dye the handle. This could mean a reduction in operating costs, since the dyeing process costs time and money.

REFERENCES

AZAMFIREI, V.; PSAROMMATIS, F.; LAGROSEN, Y. Application of automation for in-line quality inspection, a zero-defect manufacturing approach. **Journal of Manufacturing Systems**, v. 67, p. 1-22, ISSN 0278-6125, 2023.

BAHRIN, M. A. K. et al. Industry 4.0: A review on industrial automation and robotic. **Jurnal Teknologi (Sciences & Engineering)**, v. 78, n. 1-2, p. 6-13, 2016.

CHAKRABORTY, S. et al. Non-destructive testing of composite materials: A review of techniques and applications. **Composites Part B: Engineering**, v. 143, p. 172-197, 2018.

GHAEL, D. H.; SOLANKI, L.; SAHU, G. A review paper on Raspberry Pi and its applications. **International journal of advances in engineering and management**, Pilani, v. 2, n. 12, p. 225-227, 2021.

KUMAR, S. et al. A review on non-destructive testing of metallic materials for aerospace applications. **Journal of Materials Research and Technology**, v. 9, n. 4, p. 9244-9261, 2020.

LABUDZKI, R.; LEGUTKO, S.; RAOS, P. The essence and applications of machine vision. **Tehnicki Vjesnik**, v. 21, n. 4, p. 903-909, 2014.

MRUGALSKA, Beata; WYRWICKA, Magdalena K. Towards Lean Production in Industry 4.0. **Procedia Engineering**, v. 182, p. 466-473, ISSN 1877-7058, 2017.

SILVA, R.L. et al. Machine Vision Systems for Industrial Quality Control Inspections. **Product Lifecycle Management to Support Industry 4.0**, v. 540, p. 711-720, 2018.

STALLARD, M; MACKENZIE, C; CAMERON, A. A probabilistic model to estimate visual inspection error for metalcastings given different training and judgement types, environmental and human factors, and percent of defects. **Journal of Manufacturing Systems**, v. 48, p. 97-106, ISSN 0278-6125, 2018.

SZKILNYK, G. **Vision-based fault detection in assembly automation**. Thesis (Doctorate in Engineering) - Queen's University, Canada, 2012.

UPTON, E; HALFACREE, G. **Raspberry Pi User Guide**. John Wiley & Sons, 2016.

BHATTACHARYA, A.; BHATTACHARYYA, S. A Comparative Study of Programming Languages in Raspberry Pi. **International Journal of Innovative Technology and Exploring Engineering (IJITEE)**, v. 8, n. 6, p. 369-374, jun. 2019.

FONSECA, J. F.; MACEDO, D. C. Raspberry Pi applied to industrial automation using a mobile device. **IEEE Latin America Transactions**, v. 16, n. 9, p. 2372-2378, 2018.

ZHANG, L. et al. OpenCV: A computer vision library for real-time processing. **Journal of Parallel and Distributed Computing**, v. 57, p. 3-13, 2017.

BISWAS, J. et al. Industrial applications of computer vision: state-of-the-art and challenges. **Journal of Manufacturing Systems**, v. 48, p. 144-156, 2018.

SILVA, D. A. B. et al. Analysis of the advantages and disadvantages of using OpenCV in automated visual inspection. In: **Proceedings of the IX Brazilian Congress on Innovation and Product Development Management**. Sao Paulo: Brazilian Association of Product Engineering, p. 331-341, 2020.

LEVINE, M. D. **Vision in Man and Machine**. New York: McGraw-Hill College, 1985.

KAKANI V. et al. A critical review on computer vision and artificial intelligence in food industry. **Journal of Agriculture and Food Research**, v. 2, p. 4-5, 2020.

YAMADA, V. Y.; MARTINS, L. M. Industria 4.0: um comparativo da indústria brasileira perante o mundo. **Revista Terra & Cultura: Cadernos de Ensino e Pesquisa**, v. 34, n. esp., p. 95-109, 2019.

MARTINS, A. P.; JUNIOR, J. C. P. Weld inspection using computer vision. **Anais do 10° Simposio Brasileiro de Automagao Inteligente- X SBAI**, Minas Gerais, p. 444-449, 2011.

MATHE, S. E. et al. A review on raspberry pi and its robotic applications. In: **2022 2nd International Conference on Artificial Intelligence and Signal Processing (AISP)**. IEEE, 2022. p. 1-6.

APPENDIX

APPENDIX A - TESTING ALGORITHM

```
import numpy as np
import imutils
import cv2

# Defina lower and upper boundaries da cor em que você pretende trabalhar em
HSV
lower_boundary = (155, 93, 20)
upper_boundary = (179, 255, 255)

# Defina a cor do círculo de identificação do objeto em HSV
circle_color = (180, 255, 255)

camera = cv2.VideoCapture(0)

while True:
    # Read o video em tempo real
    _, frame = camera.read()

    # Você pode redefinir o tamanho do frame alterando o valor de width
    frame = imutils.resize(frame, width=800)

    # Transformando de RGB para HSV
    hsv = cv2.cvtColor(frame, cv2.COLOR_BGR2HSV)

    # Construção de uma máscara com o objetivo de performar uma série de
dilatações e erosões para remover "ruídos" da imagem
    kernel = np.ones((9, 9), np.uint8)
    mask = cv2.inRange(hsv, lower_boundary, upper_boundary)
    mask = cv2.morphologyEx(mask, cv2.MORPH_OPEN, kernel)
    mask = cv2.morphologyEx(mask, cv2.MORPH_CLOSE, kernel)

    # Encontrar os contornos do objeto na máscara criada anteriormente e
preparação para delimitação do centro do objeto
    cnts = cv2.findContours(mask.copy(), cv2.RETR_EXTERNAL,
cv2.CHAIN_APPROX_SIMPLE)[-2]
    center = None

    # Apenas prosseguir se pelo menos um contorno for identificado
    if len(cnts) > 0:
        # Encontrar o maior contorno na máscara, em seguida computar o menor
círculo circunscrito e centroide
```

```
        c = max(cnts, key=cv2.contourArea)
        ((x, y), radius) = cv2.minEnclosingCircle(c)
        M = cv2.moments(c)
        center = (int(M["m10"] / M["m00"]), int(M["m01"] / M["m00"]))

        # Apenas prossseguir, se o raio do contorno possuir um tamanho mínimo.
Ajustar conforme necessidade
        if radius > 5:
            # Desenha um círculo e adiciona uma legenda ao redor do contorno
identificado
            cv2.circle(frame, (int(x), int(y)), int(radius), circle_color, 2)
            cv2.putText(frame, "Solda", (int(x-radius), int(y - radius)),
cv2.FONT_HERSHEY_SIMPLEX, 0.6, circle_color, 2)

    # Exibir o frame
    cv2.imshow("Frame", frame)

    key = cv2.waitKey(1) & 0xFF
    # Encerra o loop ao pressionar a tecla especificada
    if key == ord("s"):
        break

camera.release()
cv2.destroyAllWindows()
```

APPENDIX B - FINAL ALGORITHM

```
import numpy as np
import imutils
import cv2
import time
from rpi_rf import RFDevice

# Defina lower and upper boundaries da cor em que você pretende trabalhar em
HSV
lower_boundary = (155, 93, 20)
upper_boundary = (179, 255, 255)

# Defina a cor do círculo de identificação do objeto
```

```
circle_color = (180, 255, 255)

# Configuração do módulo de RF
gpio_pin = 11 # GPIO que será utilizado para transmitir o sinal RF
rf_device = RFDevice(gpio_pin)
rf_device.enable_tx()

camera = cv2.VideoCapture(0)

while True:
  # Read o vídeo em tempo real
  _, frame = camera.read()

  # Você pode redefinir o tamanho do frame
  frame = imutils.resize(frame, width=800)

  # Transformando para HSV
  hsv = cv2.cvtColor(frame, cv2.COLOR_BGR2HSV)

  # Construção de uma máscara com o objetivo de performar uma série de dilatações e erosões para remover "ruídos" da imagem
  kernel = np.ones((9, 9), np.uint8)
  mask = cv2.inRange(hsv, lower_boundary, upper_boundary)
  mask = cv2.morphologyEx(mask, cv2.MORPH_OPEN, kernel)
  mask = cv2.morphologyEx(mask, cv2.MORPH_CLOSE, kernel)

  # Encontrar os contornos do objeto na máscara criada anteriormente e preparação para delimitação do centro do objeto
  cnts = cv2.findContours(mask.copy(), cv2.RETR_EXTERNAL,
```

```
cv2.CHAIN_APPROX_SIMPLE)[-2]
 center = None

 # Apenas prosseguir se pelo menos um contorno for identificado
 if len(cnts) > 0:
   # Encontrar o maior contorno na máscara, em seguida computar o menor
círculo circunscrito e centroide
   c = max(cnts, key=cv2.contourArea)
   ((x, y), radius) = cv2.minEnclosingCircle(c)
   M = cv2.moments(c)
   center = (int(M["m10"] / M["m00"]), int(M["m01"] / M["m00"]))

   # Apenas prossseguir, se o raio do contorno possuir um tamanho mínimo.
Ajustar conforme necessidade
   if radius > 50:
     # Desenha um círculo e adiciona uma legenda ao redor do contorno
identificado
     cv2.circle(frame, (int(x), int(y)), int(radius), circle_color, 2)
     cv2.putText(frame, "Solda", (int(x-radius), int(y - radius)),
cv2.FONT_HERSHEY_SIMPLEX, 0.6, circle_color, 2)
     # Transmite um sinal RF
     rf_device.tx_code(123456, protocol=1, pulselength=350)
     time.sleep(5)

 # Exibir o frame
cv2.imshow("Frame", frame)

key = cv2.waitKey(1) & 0xFF
# Encerra o loop ao pressionar a tecla especificada
if key == ord("s"):
  break

rf_device.cleanup()
camera.release()
cv2.destroyAllWindows
```

Printed by Books on Demand GmbH, Norderstedt / Germany